Study Guide

Chemistry Unit 2
for CAPE®

Roger Norris
Leroy Barrett
Annette Maynard-Alleyne
Jennifer Murray

OXFORD
UNIVERSITY PRESS

Great Clarendon Street, Oxford, OX2 6DP, United Kingdom

Oxford University Press is a department of the University of Oxford.
It furthers the University's objective of excellence in research, scholarship,
and education by publishing worldwide. Oxford is a registered trade mark of
Oxford University Press in the UK and in certain other countries

Text © Roger Norris, Leroy Barrett, Annette Maynard-Alleyne and Jennifer Murrary 2012
Original illustrations © Oxford University Press 2014

The moral rights of the authors have been asserted

First published by Nelson Thornes Ltd in 2012
This edition published by Oxford University Press in 2014

British Library Cataloguing in Publication Data
Data available

978-1-4085-1746-8

17

Printed and bound by CPI Group (UK) Ltd, Croydon, CR0 4YY

Acknowledgements

Cover photograph: Mark Lyndersay, Lyndersay Digital, Trinidad. www.lyndersaydigital.com
Illustrations: Wearset Ltd, Boldon, Tyne & Wear
Page make-up: Wearset Ltd, Boldon, Tyne & Wear

Although we have made every effort to trace and contact all
copyright holders before publication this has not been possible in all
cases. If notified, the publisher will rectify any errors or omissions at
the earliest opportunity.

Links to third party websites are provided by Oxford in good faith
and for information only. Oxford disclaims any responsibility for
the materials contained in any third party website referenced in
this work.

Contents

Introduction

This Study Guide has been developed exclusively with the Caribbean Examinations Council (CXC®) to be used as an additional resource by candidates, both in and out of school, following the Caribbean Advanced Proficiency Examination (CAPE®) programme.

It has been prepared by a team with expertise in the CAPE® syllabus, teaching and examination. The contents are designed to support learning by providing tools to help you achieve your best in CAPE® Chemistry and the features included make it easier for you to master the key concepts and requirements of the syllabus. *Do remember to refer to your syllabus for full guidance on the course requirements and examination format!*

Inside this Study Guide is an interactive CD that includes answers to the Exam-style and Revision questions and electronic activities to assist you in developing good examination techniques:

- **On Your Marks** activities provide sample examination-style short answer and essay type questions, with example candidate answers and feedback from an examiner to show where answers could be improved. These activities will build your understanding, skill level and confidence in answering examination questions.

- **Test Yourself** activities are specifically designed to provide experience of multiple-choice examination questions and helpful feedback will refer you to sections inside the study guide so that you can revise problem areas.

This unique combination of focused syllabus content and interactive examination practice will provide you with invaluable support to help you reach your full potential in CAPE® Chemistry.

1.1 Bonding in carbon compounds

Did you know?

Carbon compounds usually containing hydrogen and perhaps some other elements are often called organic compounds. About 200 years ago the Swedish Chemist J. Berzelius divided chemicals into two groups: organic and inorganic. Most organic compounds burn or char (go black) when heated. Most inorganic chemicals just melt.

The variety of carbon compounds

Carbon forms many more compounds than any other element. This is partly because, once formed, the carbon to carbon (C—C) single covalent bonds are very strong in comparison to other single covalent bonds.

Bond energies: C—C = 350 kJ mol⁻¹; N—N = 160 kJ mol⁻¹; O—O = 150 kJ mol⁻¹

It takes a lot of energy to break these strong bonds, so the compounds formed are stable. The ability of carbon atoms to form chains or ring compounds by joining together is called **catenation**. Carbon–carbon bonds are also non-polar (see *Unit 1 Study Guide*, Section 2.5) and this helps to reduce their vulnerability to attacks by other chemicals.

Hybridisation in carbon compounds

Carbon is in Group IV of the Periodic Table. It exhibits **tetravalency**. This means that it has four valence electrons in its outer principle quantum shell, which are able to form bonds with other atoms. A covalent bond is formed by the sharing of two electrons, one from each atom. Carbon can form four bonds because one of the 2s electrons in the carbon atom is transferred to a 2p orbital to give the four unpaired electrons necessary for forming four bonds (Figure 1.1.1).

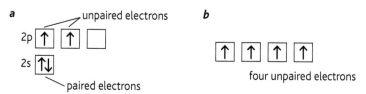

Figure 1.1.1 a *The electron configuration of carbon in the ground state;* **b** *The electron configuration of carbon when about to form covalent bonds. Each electron can form equivalent sp³ orbitals.*

The promotion of a 2s electron requires energy. But this is more than compensated for by the energy released when four bonds are formed with other C (or H, O or N) atoms. The four unfilled C atomic orbitals can be thought of as being mixed so that each has ¼ s character and ¾ p character. This process of mixing atomic orbitals is called **hybridisation**. These mixed orbitals are called sp³ hybrid orbitals. These orbitals overlap to form single bonds between carbon atoms and other carbon atoms or between carbon and H, O or N atoms. These are σ bonds (sigma bonds) (see *Unit 1 Study Guide*, Section 2.9). Figure 1.1.2 shows the formation of these bonds in ethane by the combination of separate atomic orbitals.

In ethene, one singly occupied 2s orbital and two of the three singly occupied 2p orbitals in each carbon atom hybridise to make three sp² orbitals. These have similar shapes to sp³ orbitals. These sp² orbitals form σ bonds which are arranged in a plane making a bond angle of approximately 120° with each other.

The remaining 2p orbitals from each carbon atom overlap sideways to form a π bond (Figure 1.1.3).

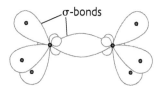

Figure 1.1.2 *The structure of ethane. The molecular orbitals formed from sp³ hybrids allow each bond to be a σ-bond.*

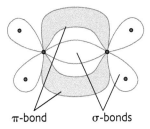

π-bond   σ-bonds

Figure 1.1.3 *Ethene has sp² orbitals in one plane making σ bonds and a π bond above and below this plane*

Resonance

In ethane and ethene the electrons are localised, i.e. they are in particular positions. In some substances, the molecular orbitals extend over three or more atoms, allowing the electrons free movement over these atoms. These electrons are said to be **delocalised**.

Benzene, C_6H_6, has six carbon atoms arranged in a ring. Figure 1.1.4(a) shows a representation of benzene. The ↔ means that the actual structure is a single (composite) form which lies between these two structures. The bonds between the carbon atoms are neither double nor single bonds. They are somewhere in-between. Making up a composite structure from several different structures is called **mesomerism**. The composite structure is called a **resonance hybrid.**

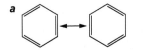

Figure 1.1.4 a *Two possible ways of representing benzene;* **b** *A modern representation of benzene*

In benzene, the six carbon atoms form a hexagon with three localised sp² hybrid orbitals (one to each hydrogen atom and two to other carbon atoms). The three sp² orbitals are arranged in a plane, so the bond angles are 120°. This leaves a single p orbital on each of the six carbon atoms. These orbitals overlap sideways to form a delocalised system of π bonds. The six electrons involved can move freely around the ring. Compounds such as benzene, which have this delocalised electron ring structure are called **aryl compounds**.

Key points

■ A large number of carbon compounds are formed by catenation – the joining of carbon atoms together to form straight or branched chains of atoms or ring compounds.

■ Most organic carbon compounds are stable because of the high value of the C—C bond energy and the non-polar nature of this bond.

■ Hybridisation of s and p atomic orbitals results in the formation of an orbital with mixed character.

■ Resonance is where the structure of a compound is a single form which is 'in-between' two or more extreme structures.

☑ *Exam tips*

It is a common error to think that resonance hybrids are mixtures of two or more forms of a structure. They are single structures. We can represent the 'in-between' structure in many cases by the use of dashed lines. Figure 1.1.5(a) shows two possible ways of representing a carboxylate ion and Figure 1.1.5(b) shows the 'in between' structure using a dashed line.

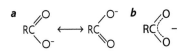

Figure 1.1.5 a *Two possible ways of representing a carboxylate ion;* **b** *The 'in-between' structure using a dashed line*

What is a homologous series?

A **homologous series** is a group of organic compounds with the same functional group in which each successive member increases by the unit $—CH_2$.

A **functional group** is an atom or group of atoms that gives a compound its particular chemical properties. For example, the functional group present in methanol, CH_3OH, and ethanol, C_2H_5OH, is $—OH$.

The table below shows the names of some homologous series and functional groups.

Homologous series	Functional group	Example
alkene	$>C=C<$	ethene, C_2H_4
alcohol	$—OH$	ethanol, C_2H_5OH
halogenoalkane	$—F, —Cl, —Br$ or $—I$	chloromethane, CH_3Cl
carboxylic acid	$-C{\overset{O}{\underset{O-H}{}}}$ or $—CO_2H$	propanoic acid, $C_2H_5CO_2H$

Each homologous series has the following characteristics:

- A particular or general formula which applies to all members in a particular series, e.g. C_nH_{2n+2} for **alkanes** (where n = number of C atoms).
- Each successive member increases by the unit $—CH_2$. For example in the alkane homologous series: CH_4, C_2H_6, C_3H_8, C_4H_{10}.
- The members have very similar chemical properties. This is because they have the same functional group.
- The physical properties change in a regular way, e.g. as the number of carbon atoms in straight-chain alkanes increases, the boiling point increases regularly.

Empirical and molecular formulae

The **empirical formula** shows the simplest whole number ratio of atoms of each element present in a compound.

The **molecular formula** shows the actual number of atoms of each element present in a molecule of a compound.

The empirical formula is sometimes the same as the molecular formula. Some examples are given in the table below:

Compound	Empirical formula	Molecular formula
ethane	CH_3	C_2H_6
pentane	C_5H_{12}	C_5H_{12}
ethanedioic acid	CO_2H	$C_2O_4H_2$
benzene	CH	C_6H_6
dinitrobenzene	$C_3H_2NO_2$	$C_6H_4N_2O_4$

Structural formulae

The **structural formula** shows the arrangement of atoms in a molecule in a simplified form. A structural formula can be either:

- a **displayed formula**, showing all atoms and bonds
- a **condensed formula**, where bonds are not shown.

For example:

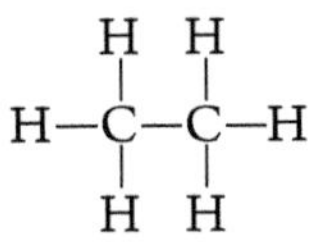

$$H-\overset{\overset{\displaystyle H}{|}}{\underset{\underset{\displaystyle H}{|}}{C}}-\overset{\overset{\displaystyle H}{|}}{\underset{\underset{\displaystyle H}{|}}{C}}-H$$

ethane (displayed)

CH_3CH_3

ethane (condensed)

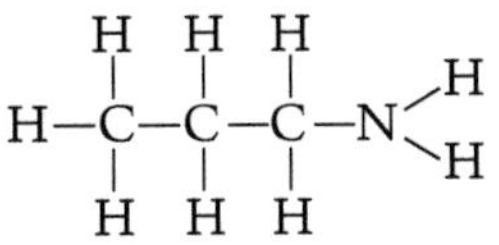

propylamine (displayed)

$CH_3CH_2CH_2NH_2$

propylamine (condensed)

With chain **hydrocarbons**, we can condense the formula even more and still show the actual structure. For example:

Hexane, $CH_3CH_2CH_2CH_2CH_2CH_3$, can be written $CH_3(CH_2)_4CH_3$

In condensed formulae, side branches coming off the main chain are shown in brackets.

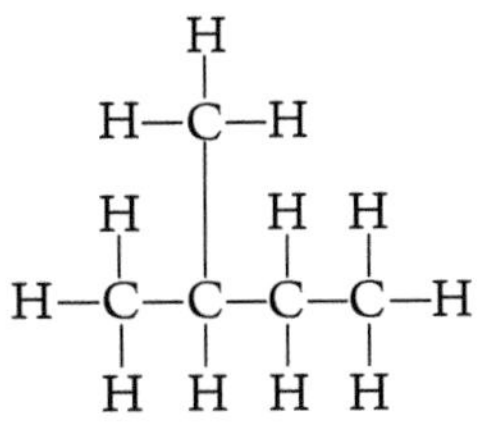

methylbutane (displayed)

$CH_3CH(CH_3)CH_2CH_3$

methylbutane (condensed)

The structure of ring compounds

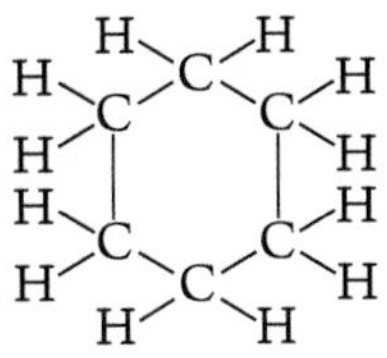

cyclohexane (displayed)

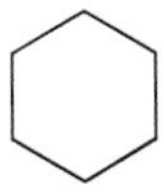

cyclohexane (condensed)

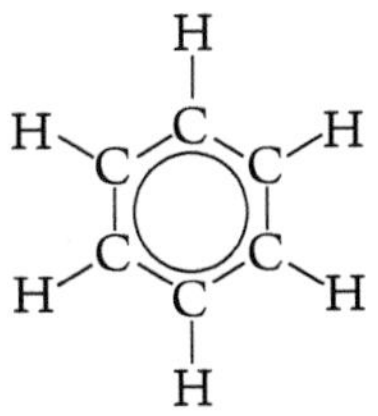

benzene (displayed)

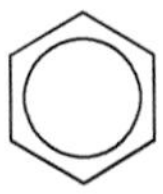

benzene (condensed)

The ring inside the hexagon in benzene represents the delocalised ring of electrons (see *Unit 1 Study Guide*, Section 2.9).

Did you know?

More than 100 000 new compounds are made each year by research chemists. Most of these are organic compounds containing rings and branches (side chains). Some metals can also be incorporated into organic compounds.

Key points

- A homologous series is a group of organic compounds with the same functional group in which each successive member increases by the unit —CH_2.

- The empirical formula shows the simplest whole number ratio of atoms of each element present in a compound.

- The molecular formula shows the actual number of atoms of each element present in a molecule of a compound.

- The structure of organic compounds can be written as displayed or condensed formulae.

On completion of this section, you should be able to:

- deduce empirical formulae using absolute masses or relative masses of elements in a compound
- deduce molecular formulae from empirical formulae
- deduce molecular formulae from combustion data.

Deducing the empirical formula

Worked example 1

In this example, we are given information about percentage (%) by mass.

Calculate the empirical formula of a compound of carbon, hydrogen and iodine that contains 8.45% carbon, 2.11% hydrogen and 89.44% iodine by mass. (A_r values: C = 12.0, H = 1.0, I = 127.0).

Step 1: Assume that we have 100 g of the compound, then each of the percentages can be converted to mass, that is 8.45 g carbon, 2.11 g hydrogen and 89.44 g of iodine.

Divide mass by A_r to determine number of moles of each atom in the compound:

$$
\begin{array}{ccc}
\text{C} & \text{H} & \text{I} \\
\dfrac{8.45}{12.0} = 0.704\,\text{mol} & \dfrac{2.11}{1.0} = 2.11\,\text{mol} & \dfrac{89.4}{127.0} = 0.704\,\text{mol}
\end{array}
$$

Step 2: Divide by lowest number to get mole ratio:

$$
\dfrac{0.704}{0.704} = 1 \qquad \dfrac{2.11}{0.704} = 3 \qquad \dfrac{0.704}{0.704} = 1
$$

Step 3: Write the formula showing the simplest ratio: CH_3I

Deducing the molecular formula

We can determine the molecular formula if we know:

- the empirical formula
- the molar mass of the compound.

The molar mass of a compound can be found by:

- weighing a known volume of gas or vapour (see *Unit 1 Study Guide*, Section 5.3)
- using a mass spectrometer (see Section 9.7).

Worked example 2

6.00 g of a hydrocarbon contains 4.80 g of carbon and 1.20 g of hydrogen. The relative molecular mass of the hydrocarbon is 30.

First work out the empirical formula:

Step 1: Divide mass by A_r:

$$
\begin{array}{cc}
\text{C} & \text{H} \\
\dfrac{4.80}{12.0} = 0.400\,\text{mol} & \dfrac{1.2}{1.0} = 1.2\,\text{mol}
\end{array}
$$

Step 2: Divide by lowest number to get mole ratio:

$$
\dfrac{0.400}{0.400} = 1 \qquad \dfrac{1.20}{0.400} = 3
$$

Step 3: Write the formula showing the simplest ratio of atoms: CH_3

Then deduce the molecular formula:

Step 4: Find the empirical formula mass: $12.0 + (3 \times 1.0) = 15.0$

Step 5: Divide the molar mass of the compound by the empirical formula mass:

$$\frac{30}{15} = 2$$

Step 6: Multiply each atom in the empirical formula by the number deduced in Step 5: $CH_3 \times 2 = C_2H_6$

Molecular formula using Avogadro's law

The worked example below shows how we can use Avogadro's law to deduce the molecular formula of a compound using combustion data. Avogadro's law states that equal volumes of all gases at the same temperature and pressure have equal numbers of molecules.

Worked example 3

Propane contains carbon and hydrogen only. When $25\,cm^3$ of propane reacts with exactly $125\,cm^3$ oxygen, $75\,cm^3$ of carbon dioxide is formed. Deduce the molecular formula of propane and write a balanced equation for the reaction.

Step 1: Write the information below the unbalanced equation:

$$C_xH_y(g) \; + \; O_2(g) \; \rightarrow CO_2(g) + H_2O(l)$$
$$25\,cm^3 \qquad 125\,cm^3 \quad 75\,cm^3$$

Step 2: Find the simplest ratio of gases and use Avogadro's law:

$$C_xH_y(g) \; + \; 5O_2(g) \; \rightarrow \; 3CO_2(g) \; + H_2O(l)$$
$$1 \text{ volume} \quad 5 \text{ volumes} \quad 3 \text{ volumes}$$

Step 3: Deduce number of C atoms: $1\,mol\,C_xH_y \rightarrow 3\,mol\,CO_2$
(so x must be 3)
$$C_3H_y(g) + 5O_2(g) \rightarrow 3CO_2(g) + H_2O(l)$$

Step 4: Deduce the number of H atoms:
$$C_3H_y(g) + 5O_2(g) \rightarrow 3CO_2(g) + H_2O(l)$$

- 6 of the 10 oxygen atoms react with carbon.
- So 4 oxygen atoms must react with hydrogen to form water.
- 4 moles of water are formed containing 8 hydrogen atoms which come from the propane.
$$C_3H_8(g) + 5O_2(g) \rightarrow 3CO_2(g) + 4H_2O(l)$$

Step 5: Write the molecular formula: $x = 3$, $y = 8$. So formula is C_3H_8

Key points

- Empirical formulae are deduced using masses or relative masses of the elements present in a compound.
- Molecular masses are found by weighing known volumes of gases or by using a mass spectrometer.
- A molecular formula can be deduced from the empirical formula if the relative molecular mass of the compound is known.
- Molecular formulae can be deduced from combustion data by applying Avogadro's law.

Naming organic compounds

On completion of this section, you should be able to:

- understand the rules for naming organic carbon compounds
- name alkanes with branched chains of carbon atoms.

The IUPAC rules

We use a set of rules to name compounds. Naming compounds in a particular way is called a systematic nomenclature. Systematic names can be used to tell us about the structure of organic carbon compounds.

Naming simple carbon compounds

Simple carbon compounds may have several parts to their name:

- *The stem:* this tells us how many carbon atoms there are along the main chain of a compound. The names of the first 10 stems are shown in the table below.

No. of C atoms	1	2	3	4	5	6	7	8	9	10
Stem	meth-	eth-	prop-	but-	pent-	hex-	hept-	oct-	non-	dec-

Did you know?

The system of rules for naming chemical compounds was drawn up by the International Union of Pure and Applied Chemistry (IUPAC) over a number of years. The task was begun by a committee of chemists in Geneva in 1892. The IUPAC was founded in 1919 by a group of chemists from industry and from universities in order to define standards of naming chemicals and measurements in chemistry which can be applied throughout the world.

- *A suffix:* this is often added to the end of the stem. This tells us about the functional groups present. For example, the suffix –ol in the name propanol tells us that the compound is in the alcohol homologous series, and the stem, prop- tells us it has three carbon atoms.
- *A prefix:* for some homologous series, the functional group appears as a prefix before the stem. For example, the bromo- in the name bromobutanol tells us that the compound is in the halogenoalkane homologous series and the but- tells us it has four carbon atoms.

Some examples are shown in the table below.

Homologous series	Suffix	No. of C atoms	Name and formula
alkane	-ane	5	pentane, C_5H_{12}
alkene	-ene	3	propene, C_3H_6
alcohol	-ol	7	heptanol, $C_7H_{15}OH$
carboxylic acid	-oic acid	1	methanoic acid, HCO_2H
ketone	-one	4	butanone, $CH_3CH_2COCH_3$

Naming branched-chain alkanes

- The position of side chains or functional groups is shown by numbering the carbon atoms.
- The longest possible chain of carbon atoms is chosen.
- Numbering starts at the end that gives the smallest number possible for the side chain.
- The side chain prefixes (comes before) the stem name.
- The side chain is named according to the number of carbon atoms it contains. For example CH_3— is methyl, C_2H_5— is ethyl, C_3H_7— is propyl. These groups are called alkyl groups. The alkyl group name is formed by changing the 'ane' of 'alkane' to 'yl'.

Example 1　$\overset{1}{C}H_3\overset{2}{C}H\overset{3}{C}H_2\overset{4}{C}H_3$　is 2–methylpentane

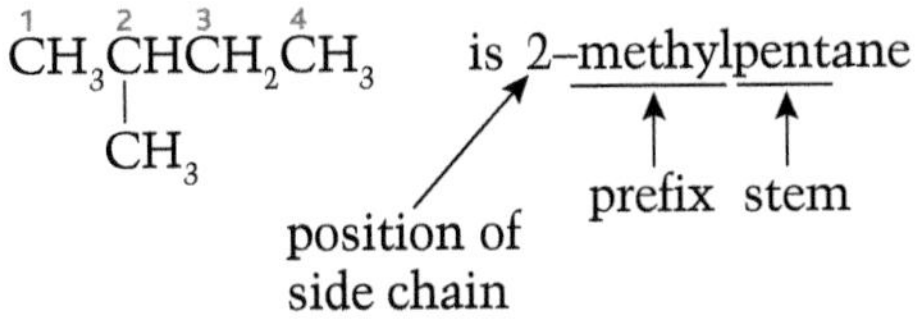

$\quad\quad\quad\quad\quad |$
$\quad\quad\quad\quad CH_3$

position of side chain　prefix　stem

Example 2　$\overset{6}{C}H_3\overset{5}{C}H_2\overset{4}{C}H_2\overset{3}{C}H\overset{2}{C}H_2\overset{1}{C}H_3$　is 3–ethylhexane

$\quad\quad\quad\quad\quad\quad |$
$\quad\quad\quad\quad\quad C_2H_5$

prefix　stem

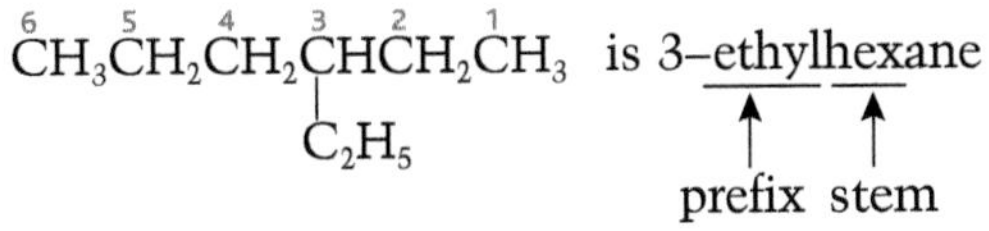

More than one side chain?

If there is more than one of the same alkyl side chain or functional group we use the prefixes di- for two groups the same, tri- for three the same and tetra- for four the same.

Example 3

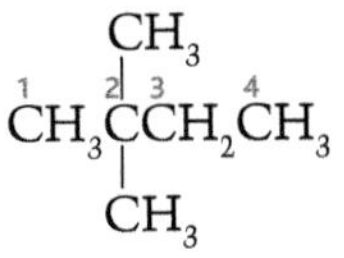

$\quad\quad\quad\quad CH_3$
$\quad\quad\quad\quad |$
$\overset{1}{C}H_3\overset{2}{C}\overset{3}{C}H_2\overset{4}{C}H_3$
$\quad\quad\quad |$
$\quad\quad\quad CH_3$　　2,2-dimethylbutane

Note:

- numbers are separated from each other by commas
- numbers are separated from words by hyphens.

If there are different side chains, they are listed in alphabetical order.

Example 4

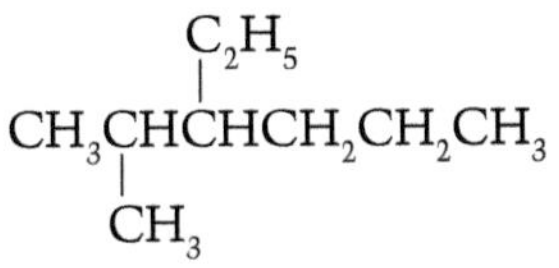

$\quad\quad\quad C_2H_5$
$\quad\quad\quad |$
$CH_3CHCHCH_2CH_2CH_3$
$\quad\quad |$
$\quad\quad CH_3$

3-ethyl-2-methylhexane

Example 5

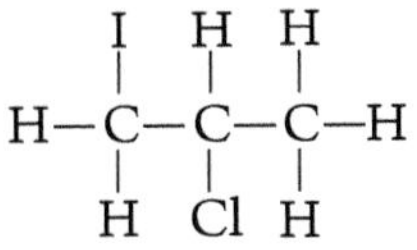

$\quad\quad I\quad\ H\quad H$
$\quad\quad |\quad\ |\quad\ |$
$H-C-C-C-H$
$\quad\quad |\quad\ |\quad\ |$
$\quad\quad H\quad Cl\quad H$

2-chloro-1-iodopropane

Functional group positions

The numbering of functional groups along the side chain follows many of the general rules. But note that the number given to the $C=C$ bond in alkenes is between the prefix and the stem.

Example 6

$CH_3CH=CHCH_2CH_3$
pent-2-ene

Example 7

$HOCH_2CH_2CH_2OH$
propane-1,3-diol

> ☑ *Exam tips*
>
> Make sure that you work out which is the longest chain carefully from 'squared off' diagrams. Remember that C—C bonds rotate freely. In the example below the longest chain is six carbon atoms.
>
> $$-C-C-C-C-C-$$
> $$\quad\ -C-$$
> $$\quad\ -C-$$

Key points

- The rules for naming organic carbon compounds are based on the use a stem, e.g. meth-, eth-, prop-, etc.
- Suffixes, e.g. –ane, -ene, -ol, are added to the stem to show the functional group present.
- Prefixes, e.g. chloro-, may be added to the stem to show the functional group present.
- Numbers are used to show the position of particular side chains.

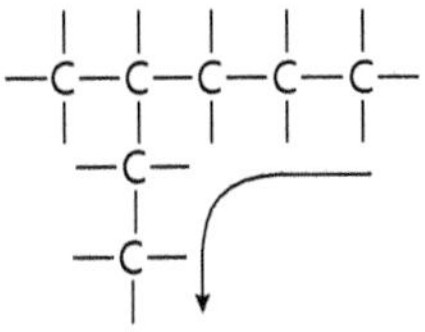

Isomers are molecules that have the same molecular formula but the atoms are arranged differently. The two main types of isomerism are:

- structural isomerism
- stereoisomerism (see Section 1.6).

Structural isomerism

Structural isomers are compounds with the same molecular formula but different structural formulae. There are three types of structural isomerism:

- chain isomerism
- functional group isomerism
- positional isomerism.

Chain isomerism

Chain isomerism is where the isomers differ in the arrangement of the carbon atoms in their carbon skeleton.

Example 1

Butane and methylpropane both have the molecular formula C_4H_{10}.

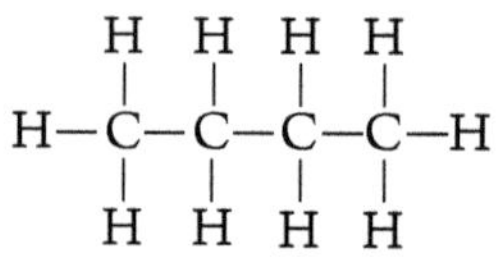

butane 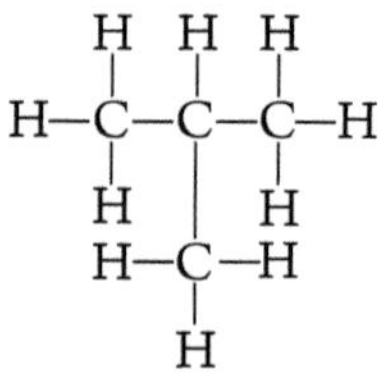   methylpropane

Note: Methylpropane is not named 2-methylpropane because there is only one possible position for the side chain.

Example 2

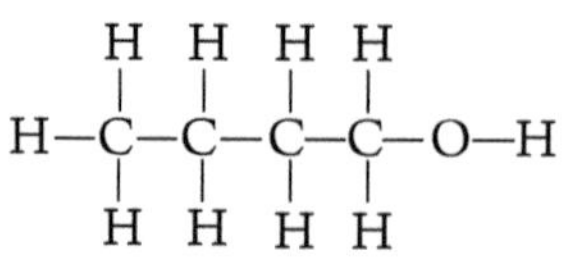

butan-1-ol 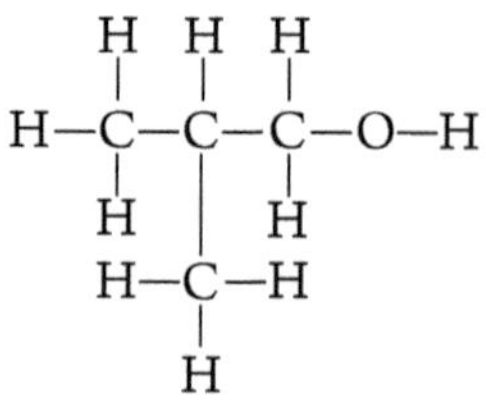   2-methylpropan-1-ol

Note: The functional group is in the 1-position, so this is not positional isomerism.

Functional group isomerism

Functional group isomerism is where the molecular formula of the isomers is the same but the functional groups are different.

Example 3 For C_2H_6O we can draw an isomer with an —OH functional group (an alcohol) and an isomer with an —O— functional group (an ether). These two isomers have different chemical and physical properties because they belong to different homologous series.

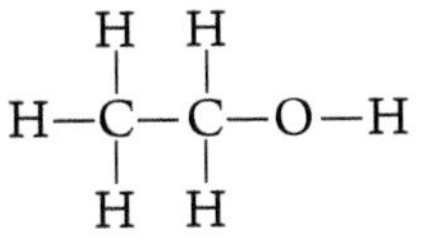

ethanol

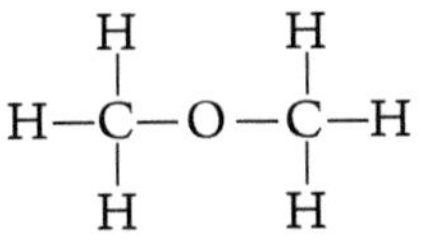

methoxymethane

Positional isomerism

Positional isomerism is where the position of the functional group is different in each isomer. The compound with the molecular formula $C_3H_6Cl_2$ has four possible isomers.

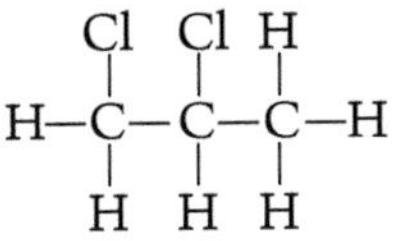

1,2-dichloropropane

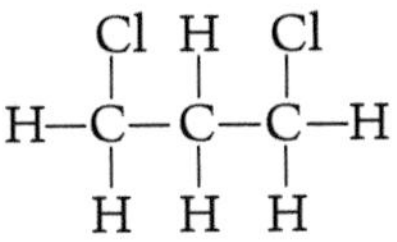

1,3-dichloropropane

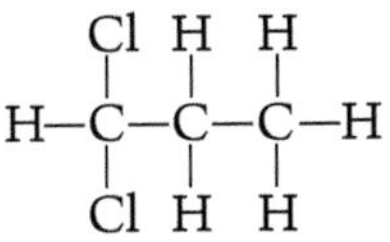

1,1-dichloropropane

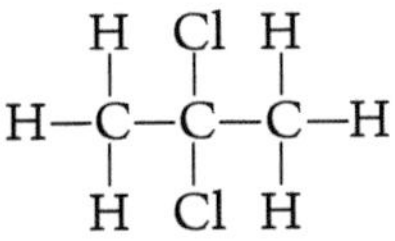

2,2-dichloropropane

Isomerism and bond rotation

There is free rotation about single bonds. Because of this, you need to take care when drawing structural formulae of different isomers, making sure that you don't repeat the same structure. For example, the two formulae below are not isomers. They are the same compound.

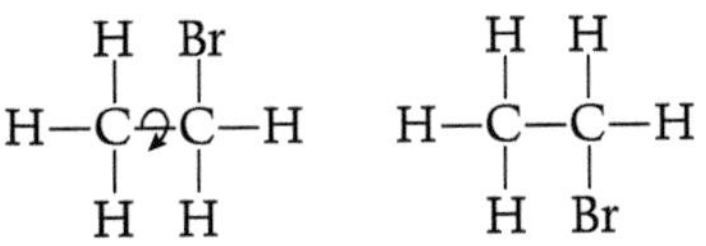

Key points

- Structural isomers are compounds with the same molecular formula but different structural formulae.

- Chain isomerism is where the isomers differ in the arrangement of the carbon atoms in their carbon skeleton.

- Functional group isomerism is where the molecular formula of the isomers is the same but the functional groups are different.

- Positional isomerism is where the position of the functional group is different in each isomer.

- Know that there is free rotation about single bonds in chains of carbon atoms.

Learning outcomes

On completion of this section, you should be able to:

- explain stereoisomerism in terms of the structure of molecules
- understand the importance of the double bond in geometrical isomerism
- know that optical isomerism is due to a particular type of asymmetry in molecules.

What is stereoisomerism?

Stereoisomerism is where two (or more) compounds have the same atoms bonded to each other but the atoms have a different arrangement in space. There are two types of stereoisomerism:

- geometrical isomerism (also called *cis-trans* isomerism)
- optical isomerism.

Geometrical (*cis-trans*) isomerism

There is free rotation around single bonds. But there is no free rotation about a $C=C$ double bond (or other double bonds). This can result in **geometrical isomerism**. Geometrical isomerism occurs when the substituent groups either side of a double bond are arranged either on the same side (*cis*) or on the opposite sides (*trans*).

Example 1

The two forms of *cis-* and *trans-*dichloroethene are different isomers.

$$\underset{\text{cis-dichloroethene}}{\overset{Cl}{\underset{H}{\diagdown}}C=C\overset{Cl}{\underset{H}{\diagup}}} \qquad \underset{\text{trans-dichloroethene}}{\overset{Cl}{\underset{H}{\diagdown}}C=C\overset{H}{\underset{Cl}{\diagup}}}$$

- In the *cis-*isomer both Cl atoms are on the same side of the $C=C$ bond.
- In the *trans-*isomer the Cl atoms are on opposite sides of the $C=C$ bond.
- The two geometric isomers have different physical properties and they may have some chemical properties which are slightly different.

Example 2

$$\underset{\text{cis-pent-2-ene}}{\overset{CH_3}{\underset{H}{\diagdown}}C=C\overset{CH_2CH_3}{\underset{H}{\diagup}}} \qquad \underset{\text{trans-pent-2-ene}}{\overset{CH_3}{\underset{H}{\diagdown}}C=C\overset{H}{\underset{CH_2CH_3}{\diagup}}}$$

Optical isomerism

Optical isomerism happens when four different groups are attached to a central carbon atom. The two isomers formed are mirror images of each other. They are not identical because they cannot be superimposed (matched up exactly) on one another. However you try to rotate them, they do not match up exactly. An example is bromochlorofluoromethane. This has four different groups attached to the central carbon and exists as two mirror images (Figure 1.6.1).

These optical isomers rotate plane-polarised light in opposite directions but by an equal amount. The electromagnetic field in plane-polarised light only vibrates in one plane unlike ordinary light which vibrates in all directions. We use an instrument called a polarimeter to measure the

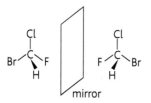

Figure 1.6.1 *The two optical isomers of bromochlorofluoromethane are mirror images*

rotation of this plane-polarised light by optical isomers. The isomer which rotates plane-polarised light in a clockwise direction is called the (+) enantiomer. The isomer which rotated it in an anticlockwise direction is called the (–) enantiomer. (Enantiomer is another word for optical isomer.)

Did you know?

The amino acids and carbohydrates in our bodies are particular forms of optical isomers. Our bodies cannot deal with their mirror images. It is fortunate for us that the amino acids and carbohydrates we get from our food are the correct optical isomers for our bodies.

Two more examples of optical isomers are shown below. You can see that the central atom does not have to be carbon as long as it has 4 different groups attached to it.

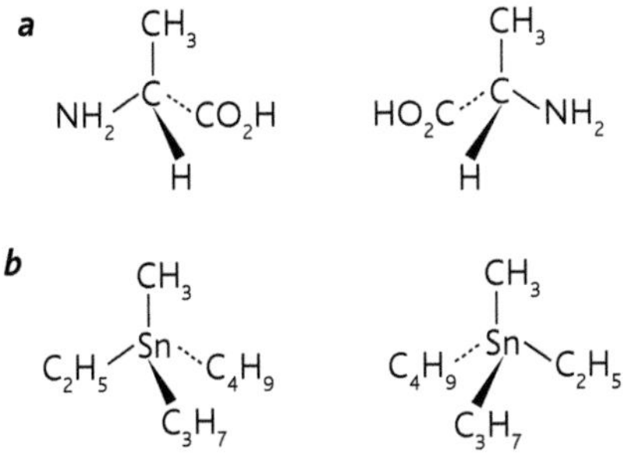

Figure 1.6.2 Optical isomers of **a** alanine (an amino acid) and **b** tin tetraalkyl

A carbon (or other atom) with four different groups attached to it is called a **chiral centre**. Some molecules, e.g. glucose, have more than one chiral centre.

Did you know?

The word chiral comes from the ancient Greek for 'hand'. Your left hand is a mirror image of your right hand but you cannot superimpose one exactly on the other.

Key points

- Geometrical isomerism is when the substituent groups either side of a double bond are arranged either on the same side (*cis*) or on the opposite sides (*trans*).

- Stereoisomerism is where two compounds have the same atoms bonded to each other but the atoms have a different arrangement in space.

- Optical isomerism happens when four different groups are attached to a central carbon atom, resulting in a molecule that has a non-superimposable mirror image. Optical isomers rotate plane-polarised light in opposite directions.

- A chiral centre (in its most common case) is an atom that has four different groups attached to it.

On completion of this section, you should be able to:

- draw all the isomers for a given molecular formula
- draw isomers for alkenes and ring structures.

Branched-chain alkanes

As the number of carbon atoms in the longest carbon chain increases, the number of possible isomers increases rapidly. For example there are 4347 isomers of pentadecane, $C_{15}H_{32}$. You may be asked to draw all the isomers of a particular alkane. The example below shows you how to do this.

Example

Draw all the isomers of hexane, C_6H_{14}:

1 Start with the straight-chain isomer (six carbon atoms in line).
2 Draw the isomers with one fewer carbon atom in the longest chain (five carbon atoms in the chain).
3 Draw the isomers with two fewer carbon atoms in the longest chain (four carbon atoms).
4 For longer-chain alkanes continue in this way until you are sure that the isomer is not one that you have already drawn.

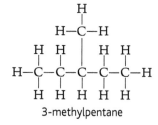

hexane

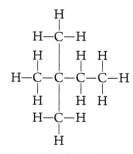

2-methylpentane

3-methylpentane

2, 2-dimethylbutane

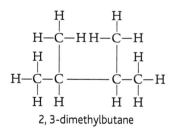

2, 3-dimethylbutane

Figure 1.7.1 *The five isomers of hexane*

When writing isomers, note that:

1 A carbon chain like this

has five carbon atoms in the longest chain, not four.

2 A carbon chain like this

is the same as this

The isomers of C_4H_8

The molecular formula C_4H_8 suggests an alkene. Possible isomers containing double bonds are:

but-1-ene

2-methylpropene

cis-but-2-ene

trans-but-2-ene

There is also the possibility of a cyclic alkene, cyclobutane.

cyclobutane

The prefix cyclo- is used to indicate a ring structure which is not an aryl compound.

Aryl compounds

Aryl compounds contain at least one benzene ring.

If a single alkyl group is attached to the ring, we do not number this group.

If there is more than one alkyl group attached to the ring we show their positions by giving them the smallest numbers possible.

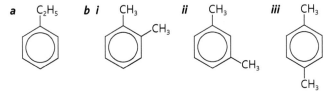

Figure 1.7.2 a Ethylbenzene; **b** The three isomers of dimethylbenzene, **i** 1,2-dimethylbenzene, **ii** 1,3-dimethylbenzene and **iii** 1,4-dimethylbenzene

Key points

- The larger the number of carbon atoms in a hydrocarbon, the greater is the number of possible isomers.
- Compounds containing rings may have two or more alkyl groups substituted in different positions in the ring.
- Alkenes may be isomeric with ring compounds.

Learning outcomes

On completion of this section, you should be able to:

- describe the functional group in the homologous series of alkanes, alkenes, alcohols, halogenoalkanes, carboxylic acids, aldehydes and ketones

- write general formulae for a given homologous series.

The alcohol homologous series

A homologous series can be identified by:

- the functional group it contains (and hence its typical chemical reactions)
- a general formula.

The **alcohol** homologous series has:

- an —OH functional group
- the general formula $C_nH_{2n+1}OH$.

The table below shows the names and structural formulae of the first ten members of the alcohol homologous series.

Number of C atoms	Name	Molecular formula	Structural formula
1	methanol	CH_4O	CH_3OH
2	ethanol	C_2H_6O	C_2H_5OH
3	propan-1-ol	C_3H_8O	C_3H_7OH
4	butan-1-ol	$C_4H_{10}O$	C_4H_9OH
5	pentan-1-ol	$C_5H_{12}O$	$C_5H_{11}OH$
6	hexan-1-ol	$C_6H_{14}O$	$C_6H_{13}OH$
7	heptan-1ol	$C_7H_{16}O$	$C_7H_{15}OH$
8	octan-1-ol	$C_8H_{18}O$	$C_8H_{17}OH$
9	nonan-1-ol	$C_9H_{20}O$	$C_9H_{19}OH$
10	decan-1-ol	$C_{10}H_{22}O$	$C_{10}H_{21}OH$

You will notice that all these alcohols, apart from methanol and ethanol, have the suffix -1-ol. This is because the —OH functional group is at the end of the chain. In alcohols larger than ethanol, the functional group can also be in other positions in the chain. If there are two —OH groups, the alcohol is a diol, if there are three —OH groups, it is a triol. For example:

propan-2-ol   pentane-2,3-diol   propane-1,2,3-triol

R groups

We can write the general formula for a homologous series by representing an alkyl group by the letter R. So for **halogenoalkanes**, $C_nH_{2n+1}X$, we can write RX. We can write different alkyl groups as R', R", R''', etc. For example, RCOR' represents the general formula of a **ketone** with two different alkyl groups.

The range of functional groups

The table below shows some examples of different functional groups you will need to know.

Homologous series	General formula	Functional group	Suffix or prefix	Example
alkanes	RH		-ane	propane, C_3H_8
alkenes	$RCH=CH_2$	$>C=C<$	-ene	propene, $CH_3CH=CH_2$
halogenoalkanes	RX (where X is a halogen)	—F —Cl —Br —I	fluoro-/ chloro-/ bromo-/ iodo-	bromoethane, C_2H_5Br
alcohols	ROH	—O—H	-ol	ethanol, C_2H_5OH
carboxylic acids	RCO_2H or RCOOH	—C(=O)O—H	-oic acid	propanoic acid, C_2H_5COOH
aldehydes	RCHO	—C(=O)H	-al	propanal, C_2H_5CHO
ketones	RCOR′	—C—C(=O)—C—	-one	propanone, CH_3COCH_3
esters	RCO_2R or RCOOR′	—C(=O)—O—C—	-oate	methyl ethanoate, CH_3COOCH_3
acyl chlorides	RCOCl	—C(=O)—Cl	-oyl chloride	ethanoyl chloride, CH_3COCl
amines	RNH_2	—N(H)(H)	-amine	methylamine, CH_3NH_2
amides	$RCONH_2$	—C(=O)—N(H)(H)	-amide	propanamide, $C_2H_5CONH_2$
arenes	C_6H_5R	R—(benzene ring)		methylbenzene, CH_3—(benzene ring)

Key points

- The functional group in alkenes is $C=C$.
- Functional groups containing oxygen are —OH (alcohols), —CHO (**aldehydes**), —CO— (ketones) and —CO_2H (carboxylic acids).
- Halogenoalkanes have the functional group —X where X is F, Cl, Br or I.
- Amines have the functional group —NH_2.
- The general formula of an organic compound can be written using R— for an alkyl group.

2 Hydrocarbons

2.1 The alkanes

Alkanes (general formula C_nH_{2n+2}), are unreactive towards most chemical reagents because there is only a very small electronegativity difference between carbon and hydrogen. They are essentially non-polar so there are no areas of higher or lower electron density that can be attacked by reagents such as acids and alkalis. Details of some important reactions of alkanes are given below.

Combustion of alkanes

Alkanes undergo combustion in excess oxygen or air to form carbon dioxide and water (in gaseous form). For example:

$$2C_4H_{10}(g) + 13O_2(g) \rightarrow 8CO_2(g) + 10H_2O(g)$$
butane

Incomplete combustion produces carbon monoxide (and/or carbon).

$$2C_4H_{10}(g) + 9O_2(g) \rightarrow 8CO(g) + 10H_2O(g)$$

Cracking of alkanes

Cracking is the thermal decomposition of alkanes into shorter-chain alkanes and alkenes. It is carried out at about 400–500°C using a catalyst of SiO_2 and Al_2O_3. A mixture of products is obtained. For example (at 400°C):

$$C_{13}H_{28}(g) \rightarrow C_8H_{18}(g) + C_3H_6(g) + C_2H_4(g)$$
tridecane　　octane　　propene　　ethene

Thermal cracking without a catalyst can also be carried out at 700–900°C. It is important because it produces the shorter-chain alkanes needed for petrol and the alkenes needed for making many chemicals, e.g. plastics.

Cracking is also a source of hydrogen:

$$CH_3CH_3(g) \rightarrow CH_2{=}CH_2(g) + H_2(g)$$

Halogenation of alkanes

When methane and chlorine are mixed in the dark there is no reaction. When methane and chlorine are mixed in the presence of ultraviolet light (or ultraviolet light from the Sun), a variety of products can be formed. This is a **substitution reaction**: A reaction in which one atom or group of atoms is replaced by another. In the chlorination of alkanes, one or more hydrogen atoms are replaced by chlorine atoms (hv represents the ultraviolet light). With excess methane:

$$CH_4(g) + Cl_2(g) \xrightarrow{hv} CH_3Cl(g) + HCl(g)$$

With excess chlorine more and more hydrogen atoms are substituted by chlorine atoms:

$$CH_3Cl(g) + Cl_2(g) \rightarrow CH_2Cl_2(l) + HCl(g)$$

$$CH_2Cl_2(l) + Cl_2(g) \rightarrow CHCl_3(l) + HCl(g)$$

$$CHCl_3(l) + Cl_2(g) \rightarrow CCl_4(l) + HCl(g)$$

Mechanism and electron movement

Reaction mechanisms show the steps in bond breaking and bond making when reactants are converted to intermediates and then to products (see *Unit 1 Study Guide*, Section 7.6).

A bond can break in two ways: homolytic and heterolytic fission.

Homolytic fission: The two shared electrons in the bond are split equally between the two atoms. Homolytic fission can occur in many types of single covalent bond.

The species formed are called free radicals. **Free radicals** are atoms or groups of atoms with unpaired electrons. The unpaired electron is represented by a dot. The movement of a single electron is shown by a fishhook arrow (see Figure 2.1.1).

Heterolytic fission: The two shared electrons in the bond are split unequally. One of the atoms keeps both pairs of electrons and so becomes negatively charged. The other atom becomes electron deficient so is positively charged. The movement of a pair of electrons is shown by a curly arrow (see Figure 2.1.2).

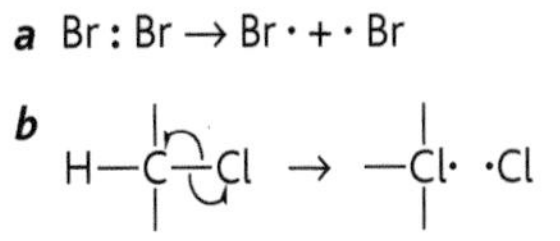

Free radical substitution in alkanes

The free radical substitution of hydrogen in alkanes by chlorine or bromine occurs in three steps, e.g. the reaction of chlorine with methane.

Initiation

The presence of ultraviolet light causes the Cl—Cl bond in chlorine to break by homolytic fission.

$$Cl{-}Cl \xrightarrow{hv} Cl{\bullet} + Cl{\bullet}$$

Propagation

Free radicals are so reactive that they can attack the relatively unreactive methane. A methyl free radical, $CH_3{\cdot}$, is formed (see Figure 2.1.3).

$$CH_4 + Cl{\bullet} \rightarrow CH_3{\bullet} + HCl$$

The methyl free radical can then attack another chlorine molecule. Chlorine free radicals are formed again. So a chain reaction occurs.

$$CH_3{\bullet} + Cl_2 \rightarrow CH_3Cl + Cl{\bullet}$$

If there is excess chlorine this process can continue until all the hydrogen atoms in the methane have been replaced.

$$CH_3Cl + Cl{\bullet} \rightarrow CH_2Cl{\bullet} + HCl$$
$$CH_2Cl{\bullet} + Cl_2 \rightarrow CH_2Cl_2 + Cl{\bullet} \text{ and so on.}$$

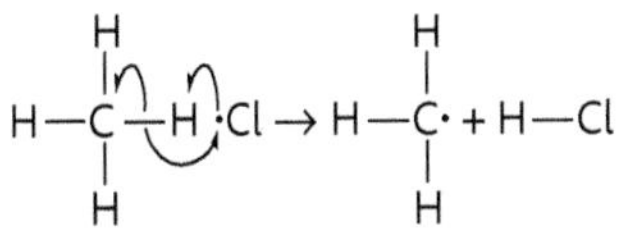

Termination

Two free radicals combine to form a single molecule (see Figure 2.1.4). For example:

$$CH_3{\bullet} + Cl{\bullet} \rightarrow CH_3Cl \quad \text{or} \quad CH_3{\bullet} + CH_3{\bullet} \rightarrow CH_3CH_3$$

This stops the chain reaction in the propagation step. The reaction finishes when there are no more free radicals left to react.

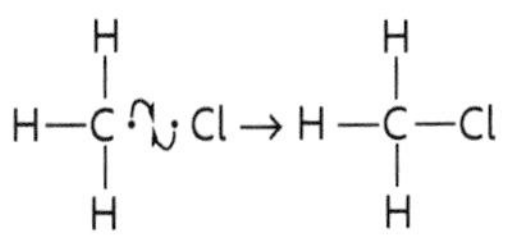

a $Br : Br \rightarrow Br{\bullet} + {\bullet} Br$

b

Figure 2.1.1 Homolytic fission: **a** The bond is split so one electron goes to each atom; **b** Fishhook arrows show the direction of movement of each electron

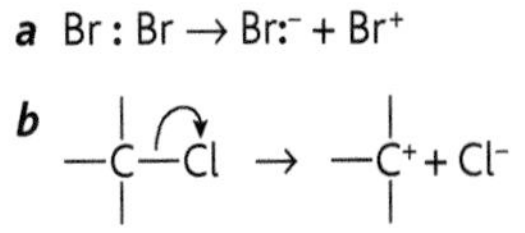

a $Br : Br \rightarrow Br{:}^- + Br^+$

b

Figure 2.1.2 Heterolytic fission: **a** The bond is split so both electrons go to one atom; **b** A curly arrow shows the direction of movement of the electron pair

Figure 2.1.3 The propogation mechanism

Figure 2.1.4 The termination mechanism

Key points

- Alkanes undergo combustion to form carbon dioxide and water.

- Cracking is the thermal decomposition of alkanes to shorter-chain alkanes and alkenes.

- Substitution reactions involve the replacement of one atom or group by another.

- There are two ways in which a bond can break: homolytic fission and heterolytic fission.

- In reaction mechanisms, movement of an electron pair is shown by a curved arrow and movement of a single electron by a fishhook arrow.

2.2 The alkenes

Learning outcomes

On completion of this section, you should be able to:

- describe the reactions of alkenes with hydrogen halides and bromine
- explain the steps involved in the mechanism of electrophilic addition of bromine and hydrogen bromide to alkenes.

Exam tips

When you draw arrows showing electron movement, remember that the tail of the curved arrow shows where an electron or electron pair moves from and the head shows where it moves to.

Alkenes have the general formula C_nH_{2n}. The structure of ethene is shown in Section 1.1. Although alkenes are non-polar, they are more reactive than alkanes. This is because the C=C double bond is an electron-rich area which can be attacked by positively charged reagents.

Electrophilic addition

An **electrophile** is a positively charged (or partially positively charged) reagent which attacks an electron-rich area of a molecule. For example H^+ ions are good electrophiles.

Most the reactions of alkenes are **addition reactions**. In addition reactions a single product is formed from two reactant molecules and no other product is made.

Addition reactions with hydrogen halides

When ethene is passed through hydrogen bromide dissolved in an inert solvent, bromoethane is formed.

$$CH_2{=}CH_2 + HBr \rightarrow CH_3{-}CH_2Br$$

The mechanism of this electrophilic addition reaction is shown in Fig 2.2.1.

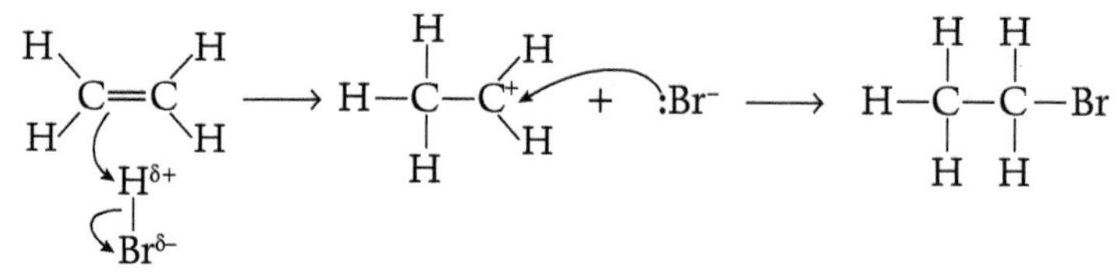

Figure 2.2.1 *The mechanism of reaction of hydrogen bromide with ethene*

- HBr is a polar molecule with a δ^+ partial charge on the H atom and δ^- partial charge on the Br atom.
- HBr acts as an electrophile attacking area of high electron density in the double bond of ethene.
- An electron pair from the double bond in ethene forms a bond with the H atom to form a positively charged **carbocation**.
- At the same time the Br atom gains control of the electron pair in the HBr to form a Br^- ion. The H—Br bond breaks heterolytically.
- The Br^- ion then attacks the + carbocation and the addition product, bromoethane, is formed.

Other hydrogen halides react in a similar way.

Exam tips

You will find the term *carbocation* useful when writing about mechanisms. A carbocation is an alkyl group carrying a single positive charge on one of its carbon atoms. They are often intermediates in organic reactions.

Reaction of HBr with other alkenes

When the double bond is not quite in the middle of the alkene a mixture of products is formed. For the reaction of HBr with propene there are two possibilities:

$$CH_2{=}CHCH_3 \quad \begin{array}{l} \xrightarrow{\text{+HBr}} CH_2Br{-}CH_2CH_3 \ \ \text{(minor product)} \\ \xrightarrow{\text{+HBr}} CH_3{-}CHBrCH_3 \ \ \text{(main product)} \end{array}$$

The rule is that the more electronegative atom (the halogen of the hydrogen halide) adds to the C atom in the alkene which is connected to the least number of H atoms.

Reaction of alkenes with bromine

Alkenes react with bromine liquid to form dibromoalkanes e.g.

$$CH_2{=}CH_2 + Br_2 \rightarrow CH_2Br{-}CH_2Br$$
$$\text{ethene} \qquad\qquad \text{1,2-dibromoethane}$$

The mechanism is similar to that for hydrogen halides (see Figure 2.2.2).

Figure 2.2.2 *The mechanism of reaction of bromine with ethene*

- ▨ The electrophile is the Br_2 molecule. As the Br_2 and ethene molecules approach each other, the high electron density in the $C{=}C$ bond repels the electron pair in the Br_2 single bond.
- ▨ This causes the Br_2 molecule to be polarised $Br^{\delta+}{-}Br^{\delta-}$ so that the $Br^{\delta+}$ end attacks the area of high electron density in the double bond.
- ▨ An electron pair from the double bond in ethene forms a bond with the $Br^{\delta+}$ and a positively charged carbocation is formed.
- ▨ At the same time the other Br atom gains control of the electron pair in the Br_2 to form a Br^- ion. The Br—Br bond breaks heterolytically.
- ▨ The Br^- ion then attacks the + carbocation and the addition product, 1,2-dibromoethane, is formed.

Chlorine reacts in a similar way.

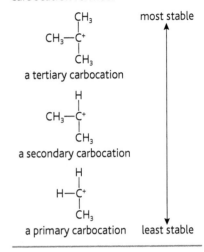
Key points

- ■ Most of the reactions of alkenes are electrophilic addition reactions.
- ■ An electrophile is a positively or partly positively charged species which attacks an area of high electron density in a molecule.
- ■ The double bond in alkenes is an area of high electron density.
- ■ The major product formed when an alkene undergoes an electrophilic addition reaction depends on the stability of the carbocation formed.

Did you know?

In addition to bromine molecules, bromine water contains bromic(I) acid, HOBr, as well as OH^- ions from the water. Bromic(I) acid is polarised $HO^{\delta-}$ —$Br^{\delta+}$ so that the $Br^{\delta+}$ end attacks the area of high-electron density in the double bond first followed by OH^- ions competing with Br^- ions (from Br_2) to attack the carbocation to form CH_2BrCH_2OH.

The bromine water test for alkenes

Liquid bromine is very hazardous. So we use aqueous bromine (bromine water) to test for the C=C bond in alkenes. Compounds containing double bonds are also described as being **unsaturated**. So the test is also a test for unsaturated compounds.

On addition of bromine water to unsaturated compounds, the colour changes from orange/red-brown (the colour of the bromine water) to colourless. The reaction is an addition reaction similar to that occurring with liquid bromine but a mixture of colourless addition products is obtained.

Reaction of alkenes with concentrated sulphuric acid

This is another example of an addition reaction. For example, ethene reacts at room temperature to form ethyl hydrogensulphate, $CH_3CH_2OSO_3H$.

$$\begin{array}{c} H \\ H \end{array}\!\!>\!\!C\!=\!C\!\!<\!\!\begin{array}{c} H \\ H \end{array} + H_2SO_4 \longrightarrow H\!-\!\overset{H}{\underset{H}{C}}\!-\!\overset{H}{\underset{OSO_3H}{C}}\!-\!H$$

The electrophile is the partially charged H atom in H_2SO_4.

$$H^{\delta+}\!\!-\!O^{\delta-}\!\!-\!SO_3H$$

When water is added to the product and the mixture warmed, ethanol (an alcohol) is formed. (The sulphuric acid is also reformed.)

$$H\!-\!\overset{H}{\underset{H}{C}}\!-\!\overset{H}{\underset{OSO_3H}{C}}\!-\!H + H_2O \longrightarrow H\!-\!\overset{H}{\underset{H}{C}}\!-\!\overset{H}{\underset{OH}{C}}\!-\!H + HOSO_3H$$

The overall reaction can be thought of as the addition of the H and OH from water across the double bond:

$$CH_2\!=\!CH_2 + H_2O \rightarrow CH_3\!-\!CH_2OH$$

Reaction with hydrogen

The addition of hydrogen to an alkene is an example of a hydrogenation reaction. It can also be regarded as addition of hydrogen or reduction.

$$CH_3CH\!=\!CH_2 + H_2 \xrightarrow[150\,°C]{Ni\ catalyst} CH_3CH_2CH_3$$
$$\text{propene} \qquad\qquad\qquad\qquad \text{propane}$$

Hydrogenation reactions are used to change vegetable oils into margarine. Hydrogenation makes the oils or fats less liquid so they have better spreading qualities. The fatty acids which are released when fats are digested may be unsaturated (they have one or more C=C double bonds). Hydrogenation also produces *trans* fatty acids (also called *trans* fats) which are not commonly found in nature (Figure 2.3.1). These *trans* fats are used in making some pies and pastries. *Trans* fats are harmful to human health. They increase levels of cholesterol in the body, leading to heart disease.

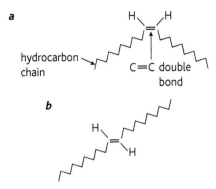

Figure 2.3.1 a A cis *fatty acid. The hydrocarbon chains are on the same side of the* C=C *bond.* **b** A trans *fatty acid. The hydrocarbon chains are on the opposite sides of the double bond.*

Reaction with potassium manganate(VII)

Potassium manganate(VII), $KMnO_4$, is a good oxidising agent. It is commonly called potassium permanganate. It is generally used as a solution acidified with sulphuric acid. Its oxidising ability depends on its concentration and the temperature used.

Cold acidified dilute potassium manganate(VII)

The purple solution turns colourless when reacted with an alkene. The alkene is converted to a diol (a compound with two —OH groups).

$$CH_2{=}CH_2 + [O] + H_2O \rightarrow CH_2OH{-}CH_2OH$$
$$\text{ethene} \qquad\qquad\qquad \text{ethane-1,2-diol}$$

This reaction can also be used to test to see if a compound is unsaturated.

Hot acidified concentrated potassium manganate(VII)

The $C{=}C$ bond in the alkene is broken and a diol is formed. The diol is then immediately oxidised by the manganate(VII) to ketones, aldehydes, carboxylic acids or carbon dioxide depending on the type of alkene. A mixture of products is often formed, for example:

$$(CH_3)_2C{=}CH_2 + 4[O] \rightarrow (CH_3)_2C{=}O + CO_2 + H_2O$$
$$\text{methylpropene} \qquad\qquad \text{propanone}$$

Note:

- We can write [O] to represent the oxidising agent, $KMnO_4$.
- You do not have to remember the equations for these reactions.

Key points

- Bromine water is used to test for the presence of C=C double bonds.
- Alcohols are formed when alkenes react with concentrated sulphuric acid and the product is then hydrolysed with water.
- Hot concentrated potassium manganate(VII) oxidises alkenes to ketones, aldehydes, carboxylic acids or carbon dioxide.
- Cold dilute potassium manganate(VII) oxidises alkenes to diols.
- The hydrogenation of fats or oils may produce *trans* fats which are harmful to health.

Revision questions

Answers to all revision questions can be found on the accompanying CD.

1 Which of the following terms best describes the reaction of hydrogen chloride with propene?
   A electrophilic addition
   B condensation
   C nucleophilic addition
   D elimination

2 Which of the statements below is correct when applied to the reaction between propene and hydrogen bromide?
   A H—Br is heterolytically cleaved.
   B Br^+ is involved in the initial attack of the propene molecule.
   C A carbanion intermediate is formed.
   D Propene undergoes hydrogenation.

3 Which of the following compounds would be expected to decolourise both bromine water and acid potassium manganate(VII) solution?
   A benzene
   B chloroethene
   C phenol
   D propane

4 Which statements are correct concerning carbon–carbon double bonds?
   i Restricted rotation about a carbon–carbon double bond prohibits the possibility of stereoisomerism.
   ii The atoms bonded to a double bond system are co-planar.
   iii Double bond systems are electron deficient.
   iv The double bond system consists of a sigma bond and a pi bond.
   A i, ii, iii
   B ii, iii
   C i, iii, iv
   D ii, iv

5 Which hydrocarbon is a member of the alkene series?
   A C_2H_2
   B C_3H_6
   C C_4H_{10}
   D C_6H_6

6 a Explain the term i 'carbocation' and ii 'electrophile', giving examples of each.
   b Use the bonding present in the alkenes to explain the nature of their chemical activity.
   c Limonene is the main contributor to the fragrance of citrus fruit, its formula is shown below:

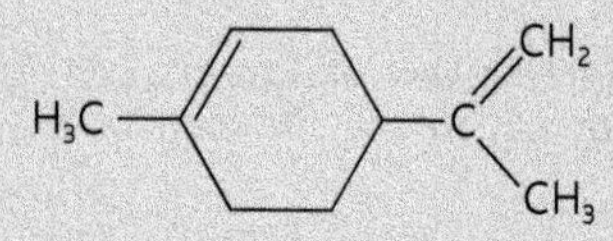

   i Suggest, giving a reason, the number of moles of bromine that will react with a limonene molecule.
   ii Write the displayed formulae of the products formed when limonene reacts with hydrogen and bromine respectively.

7 A branched chain hydrocarbon, **A**, contains 14.2% of hydrogen by mass. **A** decolourises an aqueous solution of bromine to produce two compounds **B** and **C**. In the presence of a palladium catalyst, **A** reacts with hydrogen to produce a gaseous compound **D**. **A** has a relative molecular mass of 56.
   a Calculate the empirical formula of **A** and hence its molecular formula.
   b Suggest the functional group present in **A** and give a reason.
   c Explain the mechanism involved in the production of **B** and **C** respectively, using arrows to show the movement of electrons.
   d With reference to **c** suggest which of these compounds would be preferentially formed.
   e Write the name and displayed formula for **D**.

8 a Write the displayed formulae for the compounds
A–E.

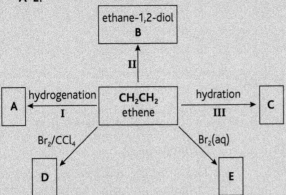

b In reactions **I–III** state the reagents and
conditions used.

c In the laboratory, reaction **III** proceeds in two
stages; write the equations which represent these
two stages.

d State two uses of **B**.

e State the commercial significance of reaction **I**.

9 10 cm³ of a gaseous hydrocarbon were mixed with
100 cm³ of oxygen and exploded. After cooling to
room temperature the resulting gaseous mixture
occupied 75 cm³. On passing the gaseous mixture
through a solution of potassium hydroxide, 30 cm³
were absorbed and the remaining gas was shown to
be oxygen. (All volumes were measured at constant
pressure.)

a Determine the molecular formula of the
hydrocarbon.

b The hydrocarbon reacts with hydrogen chloride
to produce two compounds. Give the names and
displayed formulae of these compounds.

c With reference to the respective carbocations
formed, deduce which of these compounds would
be expected to be formed in the greater quantity.

3.1 Alcohols

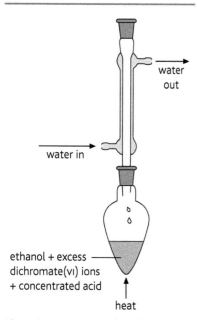

ethanol + excess dichromate(VI) ions + concentrated acid

water in

water out

heat

Figure 3.1.1 *Apparatus for refluxing primary alcohols to carboxylic acids. This allows heating without loss of ethanol, which is volatile. The ethanol vapours condense back into the flask.*

✓ Exam tips

An easy way to distinguish between primary and secondary alcohols from their structure is to remember PAL SAM. <u>P</u>rimary <u>A</u>lcohols have the —OH group at the end or <u>L</u>ast part of the chain, and <u>S</u>econdary <u>A</u>lcohols have the —OH group in the <u>M</u>iddle of the chain.

Classifying alcohols

Alcohols have the general formula $C_nH_{2n+1}OH$ or ROH. They may be classified according to the position of the —OH functional group.

Primary alcohols

e.g. propan-1-ol

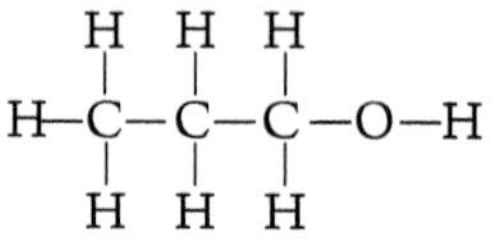

The —OH group is attached to a carbon atom that is attached to only one other carbon atom.

Secondary alcohols

e.g. propan-2-ol

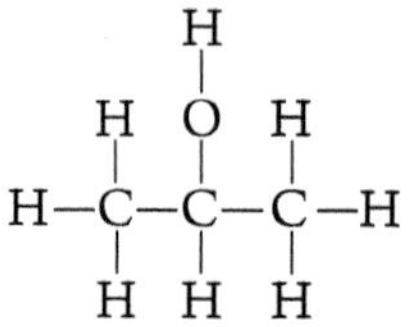

The —OH group is attached to a carbon atom that is attached to two other carbon atoms. The —OH is in the middle of the chain.

Tertiary alcohols

e.g. 2-methylpropan-2-ol

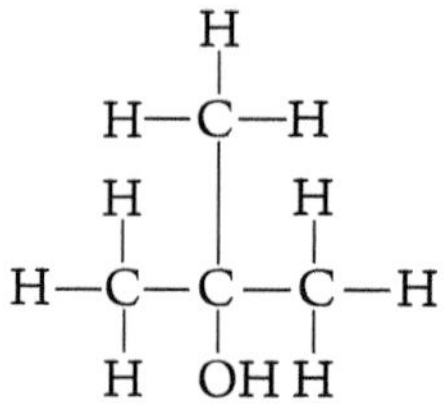

The —OH group is attached to a carbon atom that is attached to three other carbon atoms. The —OH is at a branch point in the chain.

Oxidation of alcohols

Potassium dichromate $(K_2Cr_2O_7)$, acidified with sulphuric acid is a good oxidising agent. During this reaction the orange dichromate ions are reduced to green Cr^{3+} ions.

Primary alcohols are oxidised to aldehydes when:

- the oxidising agent is not in excess and the acid is fairly dilute
- the aldehyde is distilled off immediately.

Primary alcohols are oxidised further to carboxylic acids when:

- the oxidising agent is in excess and the acid is more concentrated
- the reaction is carried out under reflux for 20 minutes (see Figure 3.1.1).

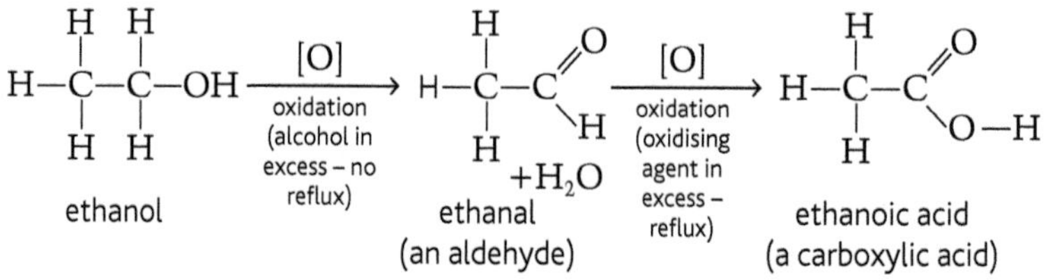

ethanol →[O] oxidation (alcohol in excess – no reflux) ethanal (an aldehyde) +H_2O →[O] oxidation (oxidising agent in excess – reflux) ethanoic acid (a carboxylic acid)

Figure 3.1.2 *Primary alcohols are converted to aldehydes at low concentrations of oxidising agent. With excess oxidising agent the aldehydes are converted to carboxylic acids.*

Secondary alcohols are oxidised to ketones. They are not oxidised further.

$$H-\overset{\overset{\displaystyle H}{|}}{\underset{\underset{\displaystyle H}{|}}{C}}-\overset{\overset{\displaystyle H}{|}}{\underset{\underset{\displaystyle OH}{|}}{C}}-\overset{\overset{\displaystyle H}{|}}{\underset{\underset{\displaystyle H}{|}}{C}}-OH + [O] \longrightarrow H-\overset{\overset{\displaystyle H}{|}}{\underset{\underset{\displaystyle H}{|}}{C}}-\overset{}{\underset{\underset{\displaystyle O}{\|}}{C}}-\overset{\overset{\displaystyle H}{|}}{\underset{\underset{\displaystyle H}{|}}{C}}-H + H_2O$$

propan-2-ol   propanone
(a ketone)

Tertiary alcohols cannot be oxidised without breaking a C—C bond in the alcohol.

Potassium manganate(VII) acts as an oxidising agent in a similar way to potassium dichromate.

Reaction of alcohols with carboxylic acids

Alcohols react with carboxylic acids to form esters. The reaction is:

- catalysed by sulphuric acid
- reversible
- carried out under reflux.

$$R-\overset{\overset{\displaystyle O}{\|\!\!\!\!\diagup}}{\underset{\underset{\displaystyle O-H}{}}{C}} + R'-O-H \overset{H^+ \text{ catalyst}}{\rightleftharpoons} R-\overset{\overset{\displaystyle O}{\|\!\!\!\!\diagup}}{\underset{\underset{\displaystyle O-R'}{}}{C}} + H_2O$$

carboxylic acid   alcohol   ester

For example:

$$C_3H_7COOH + C_2H_5OH \rightleftharpoons C_3H_7COOC_2H_5 + H_2O$$
butanoic acid   ethanol   ethyl butanoate   water

For more information about esters see Section 3.6.

Reaction with concentrated sulphuric acid

Alcohols which have at least one H atom on the C atom next but one to the —OH group react with excess concentrated sulphuric acid on heating to form alkenes.

$$CH_3CH_2CH_2OH \rightarrow CH_3CH=CH_2 + H_2O$$
propan-1-ol   propene   water

This reaction is a **dehydration** reaction: a reaction in which water is eliminated from a compound.

The iodoform reaction

Secondary alcohols, which contain the group

$$CH_3-\overset{\overset{\displaystyle OH}{|}}{\underset{\underset{\displaystyle H}{|}}{C}}-$$

are oxidised by iodine in the presence of sodium hydroxide. The triiodomethane formed precipitates as yellow crystals. This is known as the **iodoform test**. One primary alcohol, ethanol, which also contains the CH_3CHOH group also gives this reaction.

Did you know?

Sulphuric acid may also react with primary alcohols to produce ethers, especially if excess alcohol is present and the reaction is heated to no more than 140 °C.

e.g.

$$2C_2H_5OH \rightarrow C_2H_5OC_2H_5 + H_2O$$

Key points

- Primary alcohols are oxidised to aldehydes and (with excess oxidising agent) to carboxylic acids.

- Secondary alcohols are oxidised to ketones.

- Tertiary alcohols cannot be oxidised without breaking the C—C bond.

- Alcohols react with carboxylic acids to form esters.

- Hot concentrated sulphuric acid dehydrates alcohols to alkenes.

- An alkaline solution of iodine is used to test for alcohols containing the group $CH_3CH(OH)—$

Learning outcomes

On completion of this section, you should be able to:

- describe the hydrolysis of primary and tertiary halogenoalkanes
- describe nucleophilic substitution
- describe the mechanisms involved in the hydrolysis of primary and tertiary halogenoalkanes.

Classifying halogenoalkanes

Halogenoalkanes have the general formula $C_nH_{2n+1}X$ or RX (where X is a halogen). They may be classified according to the position of the halogen functional group in a similar way to alcohols. For example:

primary halogenoalkane	secondary halogenoalkane	tertiary halogenoalkane

$$H-\underset{\underset{H}{|}}{\overset{\overset{H}{|}}{C}}-\underset{\underset{H}{|}}{\overset{\overset{H}{|}}{C}}-Cl$$

$$H-\underset{\underset{H}{|}}{\overset{\overset{H}{|}}{C}}-\underset{\underset{I}{|}}{\overset{\overset{H}{|}}{C}}-\underset{\underset{H}{|}}{\overset{\overset{H}{|}}{C}}-H$$

chloroethane	2-iodopropane	2-bromo-2-methylpropane

Nucleophilic substitution

A **nucleophile** is a reagent that donates a pair of electrons to an electron-deficient atom in a molecule. A new covalent bond is formed with the electron-deficient atom.

Nucleophiles are either negatively charged ions or atoms with a partial negative charge. Examples are $:OH^-$ $:NH_3$ $:CN^-$ $H_2O:$

In a **nucleophilic substitution** reaction in halogenoalkanes:

- the nucleophile $(:Nu^-)$ replaces the halogen atom (X)
- the carbon is electron deficient because halogens are more electronegative than carbon $C^{\delta+}-Br^{\delta-}$
- the electron pair movement is from the nucleophile to the electron-deficient atom
- a bond is formed between the nucleophile and the electron deficient-carbon atom
- a bond is broken between the electron-deficient C atom and the halogen
- curly arrows show the movement of the electron pairs.

$$R-\underset{\underset{H}{|}}{\overset{\overset{H}{|}}{C}}{}^{\delta+}-X^{\delta-} \longrightarrow R-\underset{\underset{H}{|}}{\overset{\overset{H}{|}}{C}}-Nu\ +\ :X^-$$

Substitution in primary halogenoalkanes

Primary halogenoalkanes react with OH^- ions or water to form primary alcohols.

$$CH_3CH_2CH_2Cl + OH^- \rightarrow CH_3CH_2CH_2OH + Cl^-$$
$$\text{1-chloropropane} \qquad\qquad \text{propan-1-ol}$$

Hydrolysis with water is slower than with OH^- ions because water is a less effective nucleophile.

$$CH_3CH_2CH_2Cl + H_2O \rightarrow CH_3CH_2CH_2OH + H^+ + Cl^-$$

Ethanol is used as a solvent in these reactions since halogenoalkanes do not mix with aqueous solutions.

The mechanism for the reaction of bromoethane with OH⁻ ions is shown in Figure 3.2.1.

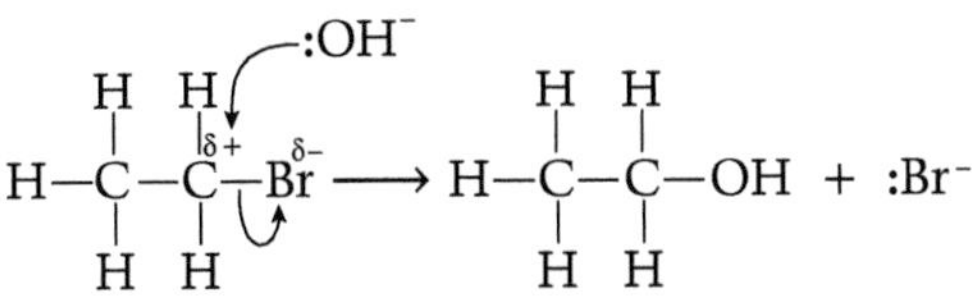

Figure 3.2.1 *Nucleophilic substitution in a primary halogenoalkane. The OH⁻ ion attacks the $C^{\delta+}$ at the same time as the C—Br bond breaks.*

In this mechanism:

- the OH⁻ ion (nucleophile) donates a pair of electrons to the $C^{\delta+}$ atom
- a new covalent bond between C and —OH is formed at the same time as the C—Br bond breaks
- the Br atom takes both electrons in the C—Br bond and leaves as a Br⁻ ion.

Substitution in tertiary halogenoalkanes

Tertiary halogenoalkanes react with OH⁻ ions or water to form tertiary alcohols.

$$CH_3CH(CH_3)ClCH_3 \quad + OH^- \rightarrow CH_3CH(CH_3)(OH)CH_3 + Cl^-$$
2-bromo-2-methylpropane 2-methylpropan-2-ol

The two-step mechanism for this reaction is shown in Figure 3.2.2.

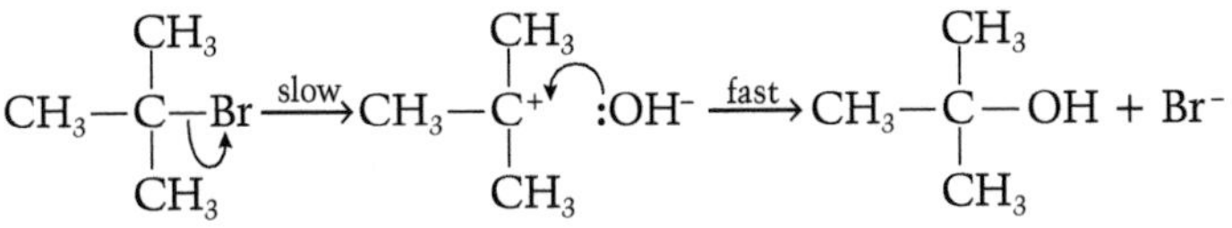

Figure 3.2.2 *Nucleophilic substitution in a tertiary halogenoalkane. In the first step a tertiary halogenoalkane ionises to form a carbocation. In the second step the OH⁻ ion attacks the carbocation.*

In this mechanism:

- the C—Br bond breaks by heterolytic fission. The bromine atom takes both electrons in the C—Br bond to become Br⁻
- an intermediate carbocation is formed with a full charge on the carbon atom
- the OH⁻ ion (nucleophile) attacks the carbocation
- a new bond is formed by the electron pair donated by the OH⁻ ion.

Chloro-, bromo- and iodoalkane hydrolysis

The reactions are similar in each case. The rate of hydrolysis is related to the bond strength of the carbon–halogen bond.

$$C_2H_5F \quad C_2H_5Cl \quad C_2H_5Br \quad C_2H_5I$$

——— faster rate of hydrolysis by OH⁻———→

——— weaker bond energy of C—halogen bond ——→

Did you know?

In the nucleophilic substitution reaction for a primary halogenoalkane, chemists think that an intermediate is formed as the C—Br bond breaks and the C—OH bond forms.

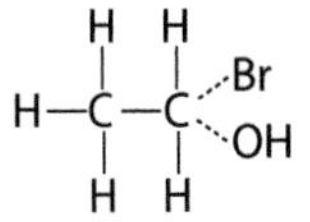

Intermediates like this have not been isolated, however.

Did you know?

The mechanism for primary halogenoalkane hydrolysis is called S_N2. (Substitution nucleophilic 2nd order). The rate-determining step involves two species (halogenoalkane and OH⁻).

The mechanism for tertiary halogenoalkane hydrolysis is called S_N1. (Substitution nucleophilic 1st order.) The rate-determining step involves one species (halogenoalkane only).

Key points

- A nucleophile is a reagent that donates a pair of electrons to an electron-deficient atom to form a new covalent bond.

- Hydrolysis of primary halogenoalkanes occurs in a single step. It involves both the halogenoalkane and OH⁻ ions.

- Hydrolysis of tertiary halogenoalkanes occurs by a two-step route which involves (i) ionisation of the halogenoalkane (ii) attack by an OH⁻ ion on a carbocation intermediate.

On completion of this section, you should be able to:

- describe the structure of aldehydes and ketones
- describe the reactions of carbonyl compounds with Brady's reagent, Tollens' reagent and Fehling's solution
- describe the reaction of carbonyl compounds with acidified potassium manganate(VII).

Did you know?

Very small amounts of propanone are present in blood and urine. People suffering from diabetes have a higher concentration than usual. Excess propanone can be exhaled through the lungs. This is called 'acetone breath'.

Structure and names of aldehydes and ketones

Aldehydes and **ketones** both contain the carbonyl group.

- In aldehydes the carbonyl carbon atom has at least one hydrogen atom bonded to it.
- In ketones the carbonyl carbon atom has two carbon atoms bonded to it.

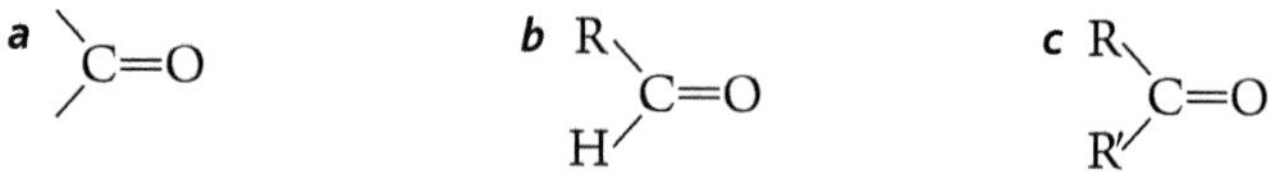

Figure 3.3.1 a The carbonyl group; **b** An aldehyde; **c** A ketone

The table below gives the names of some aldehydes and ketones.

Name of aldehyde	Structural formula	Name of ketone	Structural formula
methanal	HCHO	propanone	CH_3COCH_3
ethanal	CH_3CHO	butanone	$CH_3CH_2COCH_3$
propanal	CH_3CH_2CHO	pentan-2-one	$CH_3CH_2CH_2COCH_3$
butanal	$CH_3(CH_2)_2CHO$	pentan-3-one	$CH_3CH_2COCH_2CH_3$

Testing for the carbonyl group

We add a solution of Brady's reagent (2,4-dinitrophenylhydrazine or DNPH) to the suspected carbonyl compound. If an aldehyde or ketone is present, a deep orange precipitate is formed. If we purify the precipitate by recrystallisation and determine its melting point, we can identify the particular aldehyde or ketone present. This is because each DNPH derivative of an aldehyde or ketone has a melting point which is particular for that compound.

Distinguishing between aldehydes and ketones

Using Tollens' reagent (silver mirror test)

Tollens' reagent is an aqueous solution of silver nitrate in excess ammonia. This contains the $[Ag(NH_3)_2]^+$ ion.

Aldehydes

When warmed carefully with Tollens' reagent, aldehydes are oxidised to carboxylic acids. The silver complex ions are reduced to silver. A silver 'mirror' is seen on the side of the test-tube.

$$RCHO \quad + [O] \rightarrow RCOOH$$
$$[Ag(NH_3)_2]^+ + e^- \rightarrow \quad Ag \quad + 2NH_3$$

Note: when equations involving the oxidation of carbon compounds are complex, we can use [O] to represent the effect of the oxidising agent.

Ketones

Ketones give no reaction with Tollens' reagent (the mixture remains colourless). This is because ketones cannot be oxidised to carboxylic acids.

Using Fehling's solution

Fehling's solution is formed by mixing two solutions: Fehling's A (which contains aqueous Cu^{2+} ions) and Fehling's B (which contains a complexing reagent and an alkali).

Aldehydes

When warmed with Fehling's solution, aldehydes are oxidised to carboxylic acids. The blue colour of the Cu^{2+} ions in the Fehling's solution changes to an orange-red precipitate of copper(I) oxide.

The Cu^{2+} ions oxidise the aldehyde and are themselves reduced to the copper(I) state.

Ketones

Ketones give no reaction with Fehling's solution (the mixture remains blue). This is because ketones cannot be oxidised to carboxylic acids.

Using potassium manganate(VII) or potassium dichromate(VI)

In Section 3.1 we saw that alcohols can be oxidised by acidified potassium manganate(VII) (potassium permanganate) or potassium dichromate(VI).

In each case an aldehyde was first formed. With excess oxidising agent and using reflux, the aldehyde is converted to a carboxylic acid.

We can also use these oxidising agents to help us distinguish aldehydes from ketones.

Aldehydes

When refluxed with excess acidified potassium manganate(VII), aldehydes are oxidised to carboxylic acids. For example:

$$CH_3CH_2CHO + [O] \rightarrow CH_3CH_2COOH$$
$$\text{propanal} \qquad\qquad \text{propanoic acid}$$

The purple colour of the potassium manganate(VII) decolourises or more likely turns brownish. This is because the manganate(VII) ions are reduced to Mn^{2+} ions (very pale pink) or MnO_2 (brown).

When refluxed with excess acidified potassium dichromate(VI), the aldehyde is again oxidised to the carboxylic acid having the same number of carbon atoms. The orange colour of the potassium dichromate (oxidation number +6) changes to the green colour of Cr^{3+} ions.

Ketones

Ketones give no reaction with potassium manganate(VII) or potassium dichromate(VI). This is because ketones cannot be oxidised unless the conditions are very severe and C—C bonds are broken.

Did you know?

The tests for aldehydes using Tollens' reagent and Fehling's solution depends on the reduction of complex ions (see *Unit 1 Study Guide*, Section 13.4). Many chemical tests and some aspects of quantitative analysis by spectroscopy depend on the formation of complex ions (see Section 9.4).

Key points

- Aldehydes and ketones contain the carbonyl group, C=O.

- Aldehydes are oxidised to carboxylic acids but ketones are not.

- Brady's reagent can be used to identify a C=O group. The orange crystals formed have characteristic melting points depending on the carbonyl compound used.

- Aldehydes form a silver mirror on warming with Tollens' reagent but ketones do not.

- Aldehydes form an orange precipitate on warming with Fehling's solution but ketones do not.

On completion of this section, you should be able to:

- describe the reaction of carbonyl compounds with NaCN/HCl
- describe the mechanism of nucleophilic addition reactions of carbonyl compounds
- describe the reduction of carbonyl compounds by lithium aluminium hydride and by addition of hydrogen using a platinum catalyst.

Nucleophilic addition reactions

The $C=O$ bond in aldehydes and ketones is polarised due to the high electronegativity of the oxygen atom. The electron pairs in the bond are drawn closer to the oxygen atom.

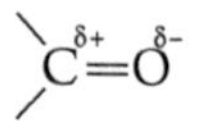

$$\overset{\delta+}{C}=\overset{\delta-}{O}$$

- The $C^{\delta+}$ atom in the carbonyl group is open to attack by nucleophiles such as $:CN^-$ and HSO_4^-.
- A negatively charged intermediate is formed.
- The intermediate reacts with a hydrogen ion (from dilute acid or water present in the reaction mixture).

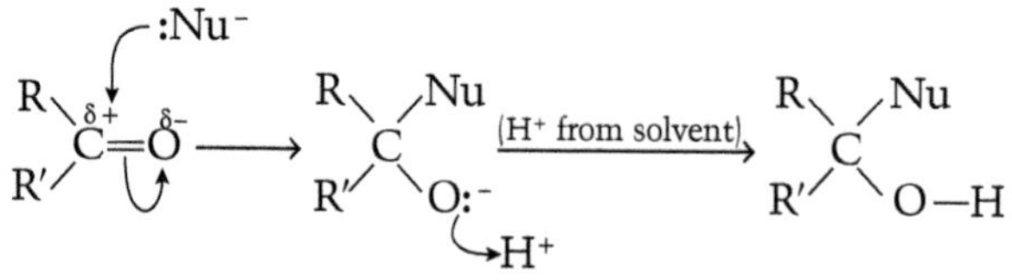

Figure 3.4.1 *A nucleophile :Nu⁻ attacks the $C^{\delta+}$ atom, a H⁺ ion from the solvent attacks the negatively charged intermediate. (Note: R' can be an alkyl group or hydrogen.)*

The overall reaction is an addition reaction because the nucleophile in the presence of hydrogen ions has added across the $C=O$ bond, e.g. with ethanal:

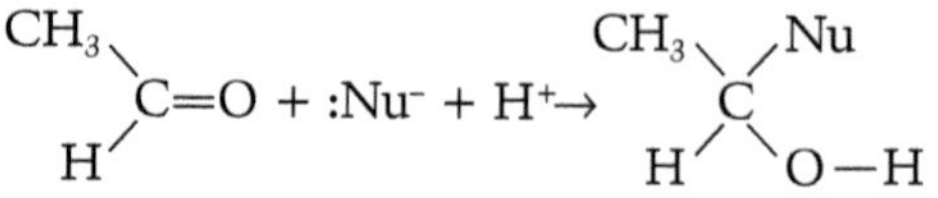

Nucleophilic addition of hydrogen cyanide

When sodium cyanide is acidified, the poisonous gas HCN is formed:

$$NaCN + H^+ \rightarrow HCN + Na^+$$

The $:CN^-$ ion in the weak acid HCN acts as a nucleophile.

The overall reaction with propanal is:

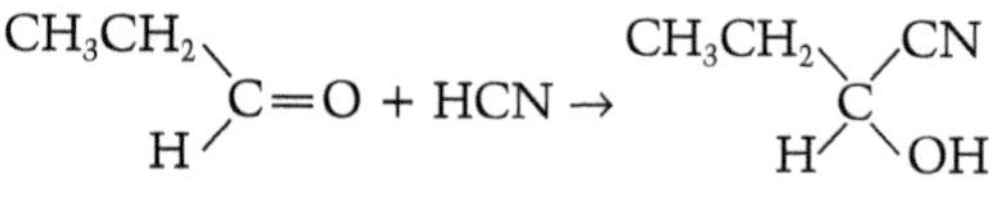

The mechanism of the reaction of hydrogen cyanide with aldehydes and ketones is shown below.

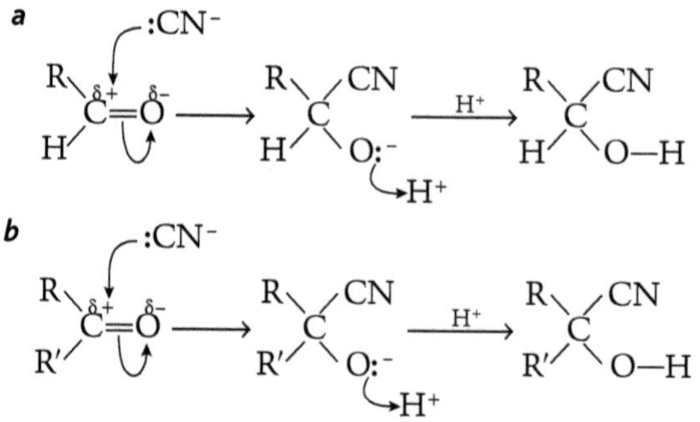

Figure 3.4.2 *Reaction of HCN with **a** an aldehyde; **b** a ketone*

In each case:

- the $C^{\delta+}$ atom in the carbonyl group is attacked by the nucleophile $:CN^-$
- a negatively charged intermediate is formed
- the intermediate reacts with a hydrogen ion (from dilute acid or water present in the reaction mixture).

Did you know?

The cyanide ion is present in plants such as sorghum, clover and cassava. It is not a danger to us, however, since it is usually present in very low concentrations.

Reduction of aldehydes and ketones

Lithium aluminium hydride (lithium tetrahydridoaluminate), $LiAlH_4$, is often used as a reducing agent in organic reactions.

- Aldehydes are reduced to primary alcohols. For example:

$$CH_3CH_2C\underset{O}{\overset{H}{<}} + 2[H] \longrightarrow CH_3CH_2\overset{H}{\underset{H}{C}}-O-H$$

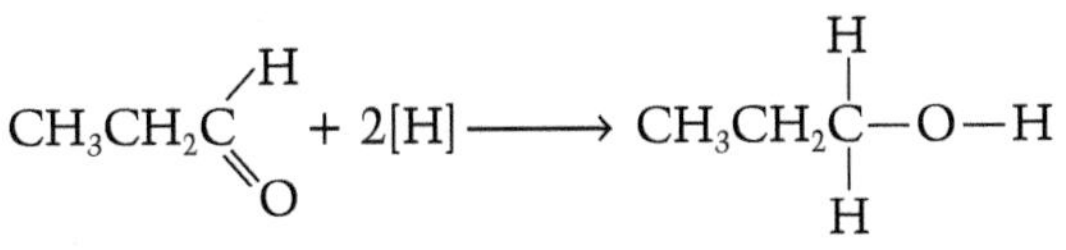

    propanal   propan-1-ol

- Ketones are reduced to secondary alcohols. For example:

$$\underset{CH_3}{\overset{CH_3}{>}}C=O + 2[H] \longrightarrow \underset{CH_3}{\overset{CH_3}{>}}\underset{H}{\overset{}{C}}-O-H$$

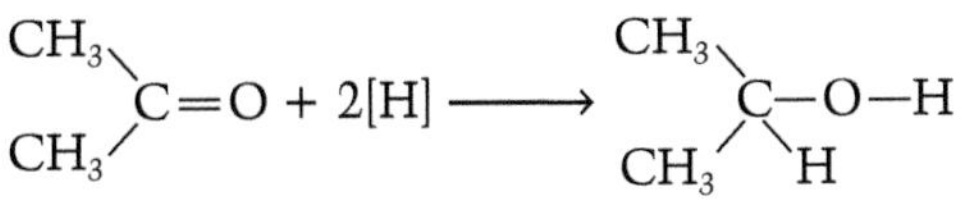

    propanone   propan-2-ol

- The reaction with $LiAlH_4$ is also a nucleophilic addition.
- The nucleophile is the hydride ion, H^-, arising from the $LiAlH_4$.
- The reaction can also be carried out by adding hydrogen in the presence of a platinum catalyst (the mechanism here is different).

Note: We can use [H] to represent the hydrogen from the reducing agent.

Key points

- Carbonyl compounds undergo nucleophilic addition reactions with the cyanide ion.
- The cyanide ion acts as a nucleophile because it donates a lone pair of electrons to the partially positively charged carbon atom in aldehydes and ketones.
- Aldehydes are reduced to primary alcohols by lithium aluminium hydride or hydrogen using a platinum catalyst.
- Ketones are reduced to secondary alcohols by lithium aluminium hydride or hydrogen using a platinum catalyst.

Introduction

The functional group in **carboxylic acids** is

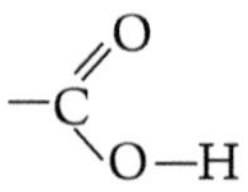

 or —CO_2H or —COOH

The table shows the names and formulae of some carboxylic acids.

Name	Structural formula
methanoic acid	HCOOH
ethanoic acid	CH_3COOH
propanoic acid	CH_3CH_2COOH
butanoic acid	$CH_3CH_2CH_2COOH$

The carboxylic acid group is polarised as shown below:

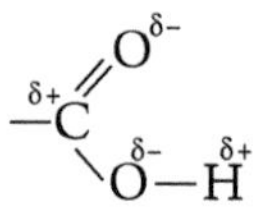

- The $C^{\delta+}$ atom can be attacked by weak nucleophiles in the presence of H^+ ions.
- The $O^{\delta-}$ atom can be attacked by positively charged species such as H^+.
- The $H^{\delta+}$ atom is lost when a carboxylic acid behaves as an acid.

Acidic properties of carboxylic acids

Carboxylic acids are weak acids. The position of equilibrium is over to the left.

$$CH_3COOH + H_2O \rightleftharpoons CH_3COO^- + H_3O^+$$
ethanoic   ethanoate
acid   ion

They are strong enough acids to show typical acid properties, e.g. reaction with sodium hydroxide, carbonates and reactive metals.

Reaction with sodium hydroxide

Carboxylic acids react with sodium hydroxide to form the sodium salt and water.

$$CH_3CH_2COOH + NaOH \rightarrow CH_3CH_2COO^-Na^+ + H_2O$$
propanoic acid   sodium propanoate

- The salts of the carboxylic acids are called carboxylates.
- $HCOO^-Na^+$ is sodium methanoate, $CH_3COO^-Na^+$ is sodium ethanoate.

Reaction with metals

Carboxylic acids react with sodium to form a metal salt and hydrogen.

$$2CH_3COOH + 2Na \rightarrow 2CH_3COO^-Na^+ + H_2$$

ethanoic acid   sodium ethanoate

Other reactive metals also form salts, e.g. magnesium ethanoate, $(CH_3COO^-)_2Mg^{2+}$

Reaction with hydrogencarbonates and carbonates

Carboxylic acids react with carbonates and hydrogencarbonates. A salt, water and carbon dioxide are formed.

$$HCOOH + NaHCO_3 \rightarrow HCOO^-Na^+ + H_2O + CO_2$$

methanoic acid   sodium methanoate

Reaction of carboxylic acids with alcohols

Carboxylic acids react with alcohols in the presence of an acid catalyst (usually concentrated sulphuric acid). The reactants are heated under reflux to prevent the loss of volatile alcohols and esters. The esters are distilled off when the reaction is complete. This type of reaction is called an **esterification** reaction.

$$\underset{\overset{\|}{O}}{CH_3C} \vdots OH + H \vdots OC_2H_5 \underset{H^+}{\rightleftharpoons} \underset{\overset{\|}{O}}{CH_3C} - OC_2H_5 + H_2O$$

Figure 3.5.1 *The ester, ethyl ethanoate, is formed by the reaction of ethanol with ethanoic acid. The dashed line shows the bonds that are broken during the reaction.*

The reaction can also be described as a **condensation reaction** – a reaction in which two molecules have reacted together and a small molecule has been eliminated. Another way of describing the reaction is as an addition–elimination reaction. This is because the ethanol molecule first forms an addition product by attacking the $C^{\delta+}$ atom of the —COOH group.

Reaction with PCl_5, PCl_3 and $SOCl_2$

Carboxylic acids react rapidly with phosphorus(III) chloride, phosphorus(V) chloride or sulphur dichloride oxide (thionyl chloride), $SOCl_2$. The products are called acid chlorides (acyl chlorides). Acidic fumes of HCl are also produced.

$$CH_3COOH + PCl_5 \rightarrow CH_3COCl + POCl_3 + HCl$$
$$CH_3COOH + SOCl_2 \rightarrow CH_3COCl + SO_2 + HCl$$

ethanoic acid   ethanoyl chloride

Safety note: PCl_3 and $SOCl_2$ should be used in a fume cupboard.

Key points

- Carboxylic acids are weak acids.
- Carboxylic acids react with sodium hydroxide to form a salt and water and with sodium hydrogencarbonate to form a salt, carbon dioxide and water.
- Carboxylic acids react with reactive metals to form a salt and hydrogen.
- Carboxylic acids react with alcohols to form esters.
- Carboxylic acids react with PCl_3, PCl_5 and $SOCl_2$ to form acid chlorides and hydrogen chloride.

 Exam tips

You do not have to learn the equations for the reactions of the phosphorus chlorides or thionyl chloride with carboxylic acids. You should know, however, that acidic fumes are given off and that acid chlorides are formed.

On completion of this section, you should be able to:

- name esters and write their structural formulae
- describe the formation of esters from alcohols
- describe the acid- and base-catalysed hydrolysis of esters.

Introduction

Esters have the general structure

R—C with =O, O—R', labelled ester link, from acid, from alcohol

The —COO— group is often called an ester link (see Section 6.2).

The naming of esters is based on the name of the carboxylic acid and alcohol used to make them.

- The name begins with the alkyl (or aryl) group from the alcohol.
- The name ends with the part coming from the carboxylic acid, but -oic acid is changed to -oate.

CH_3—C with =O, O—CH_3 = methyl ethanoate

ethanoate (from ethanoic acid)   methyl (from methanol)

The table gives some names of different esters.

Structure of ester	Name of ester
$CH_3CH_2CH_2COOCH_3$	methyl butanoate
$HCOOCH_2CH_2CH_3$	propyl methanoate
$CH_3COOCH_2CH_3$	ethyl ethanoate
⬡—$COOCH_2CH_3$	ethyl benzoate
CH_3COO—⬡	phenyl ethanoate

Hydrolysis of esters

Hydrolysis is the breakdown of a compound with water. It is often speeded up by reacting a compound with either an acid or an alkali. Esters are hydrolysed by heating the ester under reflux with an acid or a base.

- Heating is necessary because the reaction is slow.
- The acid or alkali acts as a catalyst.
- Reflux is necessary to prevent the loss of the volatile vapours of the ester and alcohol. The vapours rise and then condense on the colder parts of the condenser. They then drip back into the flask (Figure 3.6.1).

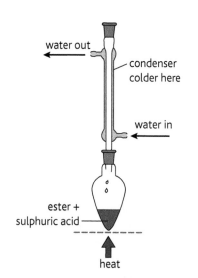

water out

condenser colder here

water in

ester + sulphuric acid

heat

Figure 3.6.1 *Acid hydrolysis of an ester by refluxing*

Acid hydrolysis

This is the reverse of the preparation of an ester from an alcohol and a carboxylic acid.

- ▪ The ester is heated with diluted sulphuric acid.
- ▪ The reaction is reversible.
- ▪ So the ester is not fully hydrolysed.
- ▪ A carboxylic acid and an alcohol are formed.
- ▪ For an ester RCOOR′ the carboxylic acid arises from the RCO— part of the ester and the alcohol from the —OR part.

$$CH_3-\overset{\displaystyle O}{\underset{\displaystyle O-C_2H_5}{C}} + H_2O \underset{H^+ \text{ catalyst}}{\rightleftharpoons} CH_3-\overset{\displaystyle O}{\underset{\displaystyle OH}{C}} + C_2H_5OH$$

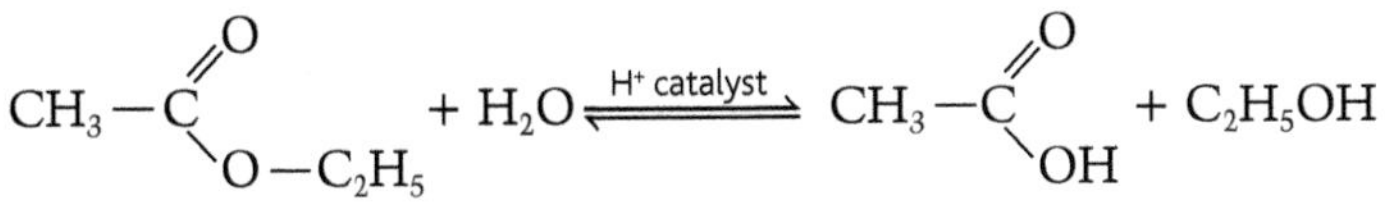

Figure 3.6.2 Acid hydrolysis of ethyl ethanoate

Alkaline hydrolysis

In the base-catalysed hydrolysis of an ester:

- ▪ The ester is heated with aqueous sodium hydroxide (or other suitable base).
- ▪ The reaction is not reversible.
- ▪ The ester is fully hydrolysed.
- ▪ An alcohol and the salt of a carboxylic acid are formed.
- ▪ For an ester RCOOR′, the salt of the carboxylic acid arises from the RCO— part of the ester and the alcohol from the —OR part.

$$CH_3-\overset{\displaystyle O}{\underset{\displaystyle O-C_2H_5}{C}} + NaOH \rightarrow CH_3\overset{\displaystyle O}{\underset{\displaystyle O^-Na^+}{C}} + C_2H_5OH$$

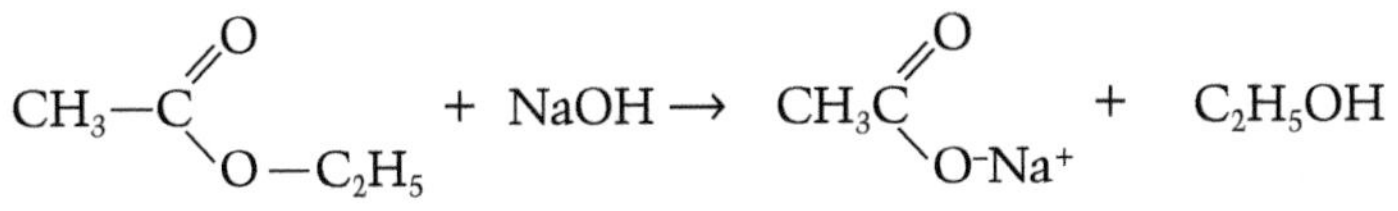

$$CH_3COOCH_2CH_3 + NaOH \rightarrow CH_3COO^-Na^+ + CH_3CH_2OH$$
ethyl ethanoate sodium ethanoate ethanol

Figure 3.6.3 Alkaline hydrolysis of ethyl ethanoate

The sodium salt of the carboxylic acid is formed because carboxylic acids react with sodium hydroxide (see Section 3.5). Alcohols do not react with sodium hydroxide.

Some more equations for ester hydrolysis

$$HCOOCH_3 + H_2O \rightleftharpoons HCOOH + CH_3OH$$
methyl methanoate   methanoic acid   methanol

$$CH_3CH_2COOCH_3 + H_2O \rightleftharpoons CH_3CH_2COOH + CH_3OH$$
methyl propanoate   propanoic acid   methanol

$$CH_3COOCH_2CH_2CH_3 + NaOH \rightarrow CH_3COO^-Na^+ + CH_3CH_2CH_2OH$$
propyl ethanaote   sodium ethanoate   propanol

$$HCOOCH_2CH_3 + NaOH \rightarrow HCOO^-Na^+ + CH_3CH_2OH$$
ethyl methanoate   sodium methanoate   ethanol

Exam tips

You may find it useful for revision purposes to draw a 'spider diagram' or 'mind map' showing all the reactions between the various functional groups in this section. This will help you to see how you can synthesise a particular compound by a three- or four-stage route starting from another particular compound. In your spider diagram you should include alkenes, alcohols, halogenoalkanes, aldehydes, ketones, carboxylic acids and esters.

Did you know?

Apart from their common uses in flavourings and perfumes, some esters can be used to reduce insect damage. Some esters act as pheromones – they are given out naturally by female insects to attract the males. Spraying crops with synthetic pheromones confuses the males and they do not find females to mate with. So fewer offspring are produced.

Key points

- ▪ Esters are formed by refluxing alcohols with acid.
- ▪ When esters are hydrolysed by acids, the products are a carboxylic acid and an alcohol.
- ▪ When esters are hydrolysed by a base, the products are the salt of a carboxylic acid and an alcohol.

Learning outcomes

On completion of this section, you should be able to:

- describe saponification
- understand that biodiesel can be made by transesterification reactions
- understand the basic principles of transesterification.

Fats and oils

- Fats and oils are esters of long-chain carboxylic acids with glycerol.
- Long-chain carboxylic acids are sometimes called fatty acids.
- The only difference between a fat and an oil is that a fat is solid and an oil is a liquid at room temperature.
- The chain lengths of the fatty acids (carboxylic acids) in fats are from 12–18 carbon atoms.
- The fatty acids in fats can be the same or different.

Figure 3.7.1 shows glycerol reacting with three fatty acid molecules to make a triglyceride by an esterification reaction.

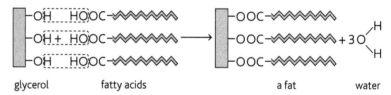

-OH HOOC-
-OH + HOOC-
-OH HOOC-
-OOC-
-OOC- + 3O
H
H
-OOC-
glycerol
fatty acids
a fat
water

Figure 3.7.1 *Glycerol reacts with up to three fatty acids to form a fat (〰 represents the alkyl side chain of the fatty acid)*

Saponification

Saponification is the process of making soaps by the hydrolysis of fats and oils.

- Soaps are metal salts of long-chain carboxylic acids.
- Soaps are made by boiling fats with sodium or potassium hydroxide.
- Sodium hydroxide hydrolyses the three ester links in fats.
- The products are the sodium salts of long-chain carboxylic acids (soap) and glycerol.

$$\text{fat (or oil)} + \text{sodium hydroxide} \xrightarrow{\text{heat}} \text{soap} + \text{glycerol}$$

Figure 3.7.2 shows the hydrolysis of a fat (trigyceride) with sodium hydroxide.

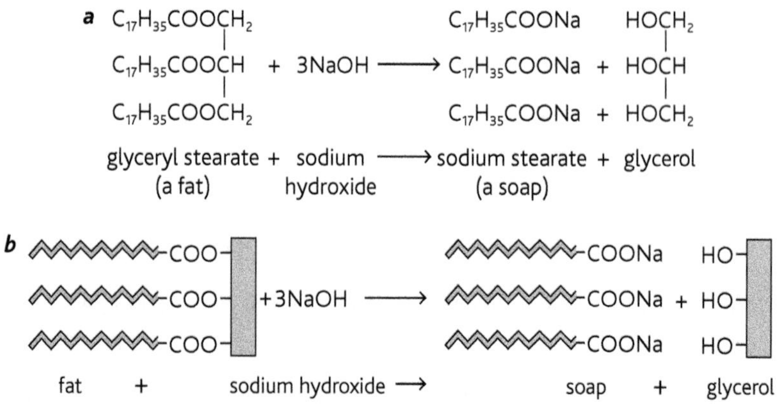

a
$C_{17}H_{35}COOCH_2$
$C_{17}H_{35}COOCH$ + 3NaOH
$C_{17}H_{35}COOCH_2$
$C_{17}H_{35}COONa$ $HOCH_2$
$C_{17}H_{35}COONa$ + HOCH
$C_{17}H_{35}COONa$ + $HOCH_2$
glyceryl stearate + sodium hydroxide → sodium stearate + glycerol
(a fat)
(a soap)
b
-COO-
-COO- +3NaOH
-COO-
-COONa HO-
-COONa + HO-
-COONa HO-
fat + sodium hydroxide → soap + glycerol

Figure 3.7.2 *Soaps are formed when fats or oils are hydrolysed with sodium hydroxide;* ***a*** *The hydrolysis of a fat;* ***b*** *A simplified diagram of saponification (〰 represents the alkyl side chain of the fatty acid)*

Biodiesel

The supply of petroleum used as a basis for fuels will eventually run out. For this reason scientists have been trying to find other ways of making fuels. **Biodiesel** is a fuel for diesel engines that is made from vegetable oils from plants or fats from animals. It is made by transesterification.

Transesterification is the reaction of an ester with an alcohol to form a different ester and different alcohol. The alkyl group from the alcohol replaces the alkyl group in the ester which originates from a different alcohol. The reaction is slow so:

- heat is required
- a catalyst is used to speed up the process. Acids, alkalis or alkoxides e.g. sodium methoxide, $CH_3O^-Na^+$ are used as catalysts.

$$C_2H_5\overset{O}{\overset{\|}{C}}OC_6H_{13} + CH_3OH \xrightarrow{\text{catalyst}} C_2H_5\overset{O}{\overset{\|}{C}}OCH_3 + C_6H_{13}OH$$

ester 1   alcohol 1   ester 2   alcohol 2

Figure 3.7.3 *Transesterification. An ester reacts with an alcohol to form a different ester and a different alcohol.*

Transesterification usually results in a simpler ester being formed from a more complex ester.

$$CH_3COOC_{12}H_{25} + CH_3OH \rightarrow CH_3COOCH_3 + C_{12}H_{25}OH$$
dodecanyl ethanoate + methanol → methyl ethanaote + dodecanol

The ester is more useful as a fuel because it has a lower molar mass. So it has a lower viscosity and burns more easily. Fats and oils can also undergo transesterification, especially in the presence of sodium hydroxide where hydrolysis also occurs.

$$\begin{array}{l} CH_2-O_2CR' \\ | \\ CH-O_2CR'' \\ | \\ CH_2-O_2CR''' \end{array} + 3CH_3OH \longrightarrow \begin{array}{l} CH_3O_2CR' \\ CH_3O_2CR'' \\ CH_3O_2CR''' \end{array} + \begin{array}{l} CH_2OH \\ | \\ CHOH \\ | \\ CH_2OH \end{array}$$

fat   glycerol

Figure 3.7.4 *Transesterification and hydrolysis of a fat. A triglyceride reacts with methanol to form esters of lower molar mass and glycerol. R, R' and R'' represent alkyl groups with 12–18 carbon atoms.*

The fat which has a high molar mass is converted to simpler esters of lower molar mass.

Key points

- Fats and oils are esters of glycerol and long-chain carboxylic acids (fatty acids).
- Soaps are made by the hydrolysis of fats or oils by boiling with sodium hydroxide.
- Biodiesel is made by the process of transesterification.
- Transesterification involves the reaction of fats or oils with an alcohol of low relative molecular mass. A different ester and glycerol are formed.

Did you know?

The cleaning action of soap is due to the attraction of different parts of the soap for grease and water. The ionic 'head' is attracted to water and the hydrocarbon 'tail' is attracted to grease.

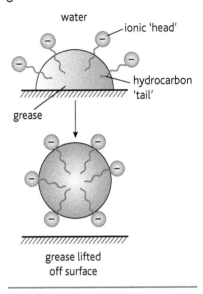

grease lifted off surface

✅ Exam tips

You do not have to remember the equations in this section. You should, however, know:

1. the basic structure of a triester (fat)

2. how the ester link is broken when fats are saponified

3. how in transesterification, the alkyl group of the alcohol swaps with the alkyl group in the —COOR group of the ester.

Testing for the C=C double bond

Aqueous bromine (bromine water) is used to test for the C=C bond in alkenes. Compounds containing double bonds are also described as being unsaturated. So the test is also a test for unsaturated compounds.

On addition of bromine water to unsaturated compounds, the colour of the bromine changes from orange to colourless.

We can distinguish alkenes from alkanes by the bromine water test. Alkanes do not give a positive result with this test.

Halogenoalkanes

Halogenoalkanes (in a solution of ethanol) react with OH^- ions form alcohols and halide ions.

$$CH_3CH_2CH_2Cl + Na^+ + OH^- \rightarrow CH_3CH_2CH_2OH + Na^+ + Cl^-$$
1-chloropropane   propan-1-ol

The halide ions produced in this reaction can be identified using silver nitrate.

- Hydrolyse the suspected halogenoalkane with sodium hydroxide.
- Add excess nitric acid to neutralise the OH^- ions.
- Add a few drops of aqueous silver nitrate.
- Observe any precipitate formed:
    - chloroalkanes give a white precipitate which darkens rapidly
    - bromoalkanes give a cream precipitate which darkens slowly
    - iodoalkanes give a light yellow precipitate which does not darken.

Aldehydes and ketones

Test for carbonyl group in aldehydes and ketones

Add a solution of Brady's reagent to the suspected carbonyl compound.

If an aldehyde or ketone is present a deep orange precipitate is formed (see Section 3.3 for more information).

Carboxylic acids do not give a positive reaction to this test because of the delocalisation in the carboxylate ion.

Distinguishing between aldehydes and ketones

We can distinguish aldehydes from ketones by three tests.

1 **Using Tollens' reagent (silver mirror test) – see Section 3.3**
    - If an aldehyde is present, a silver 'mirror' is seen on the side of the test-tube.
    - Ketones do not react (the mixture remains colourless). No silver 'mirror' is seen.
    - Methanoic acid gives a positive reaction because it has a CHO group but it can be distinguished from aldehydes because it has the acidic properties typical of carboxylic acids.

2 Using Fehling's solution – see Section 3.3

- If an aldehyde is present an orange-red precipitate of copper(I) oxide is formed.
- Ketones do not react (the mixture remains blue).
- Methanoic acid gives a positive reaction.

3 Using potassium dichromate(VI) or potassium manganate(VII) – see Section 3.3

- If an aldehyde is present the orange potassium dichromate(VI) turns green or the purple potassium manganate(VII) decolourises.
- If a ketone is present no colour change is observed on heating.
- Primary and secondary alcohols will also give a positive result in this test but they do not give a positive test with Brady's reagent.

Carboxylic acids and acid halides

We can distinguish carboxylic acids from alcohols (and many other organic compounds) by the following reactions:

1 Acidity

A solution of a carboxylic acid in water has a pH below 7.

2 Reaction with sodium hydrogencarbonate or sodium carbonate

Bubbles of carbon dioxide, CO_2, are released when the acid is added to a carbonate. The presence of CO_2 is detected by bubbling the gas through limewater. The limewater turns milky if CO_2 is present.

Acid halides can be distinguished from carboxylic acids by adding them to water. The reaction is very vigorous and highly acidic choking fumes are given off, so a fume cupboard should be used for these reactions.

Alcohols and esters

Alcohols can be distinguished from all the other homologous series above apart from carboxylic acids because they react with sodium. Hydrogen is given off (pops with a lighted splint).

Alcohols can be distinguished from carboxylic acids because:

- carboxylic acids are acidic and react with sodium hydrogencarbonate
- alcohols are not acidic so do not react with sodium hydrogencarbonate.

Alcohols which contain the group $CH_3CH(OH)$ — (many secondary alcohols and ethanol) can be distinguished by the iodoform test:

- add iodine and sodium hydroxide
- a precipitate of yellow crystals shows the presence of a secondary alcohol or ethanol.

Esters are distinguished from most other homologous series by their fruity smell.

Key points

- Different functional groups can be identified used qualitative tests involving colour changes.

- Halogenoalkanes can be distinguished by hydrolysing them with sodium hydroxide and testing the ionic product with aqueous silver nitrate.

- Alcohols, aldehydes and ketones can be identified by various oxidation reactions and by reaction with Brady's reagent.

- Carboxylic acids can be distinguished from alcohols by their acidic nature.

4 Aromatic compounds

4.1 Some reactions of benzene

Learning outcomes

On completion of this section, you should be able to:

- describe the bromination and nitration of benzene
- explain the steps involved in the mechanism for the nitration and bromination of benzene.

Arenes

Arenes are hydrocarbons based on **benzene**. Benzene has a six-membered planar ring with a delocalised system of π electrons above and below the plane of the ring (see *Unit 1 Study Guide*, Section 2.9). This stabilises the structure of benzene.

Did you know?

Some arenes have been extracted as sweet-smelling oils from plants for many centuries. This is why they were named **aromatic** (aroma-producing) compounds. Many arenes made in the laboratory do not have such nice smells! In chemistry the term *aromatic* now refers to the structure of the compounds rather than the smell.

Electrophilic substitution in benzene

The delocalised ring of π electrons in benzene has a high-electron density and is exposed to attack by electrophilic reagents. Figure 4.1.1 shows the general mechanism.

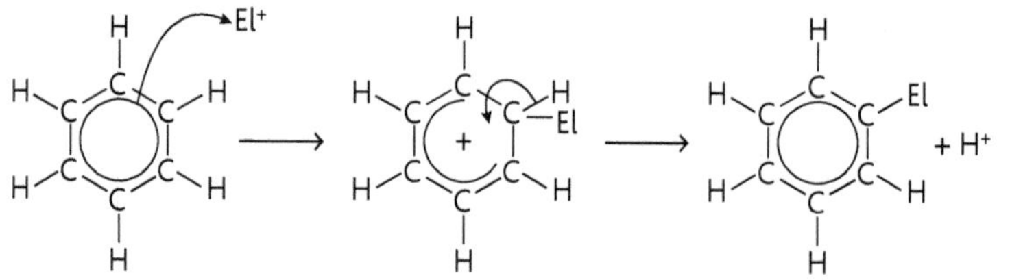

***Figure 4.1.1** General mechanism of nucleophilic substitution in benzene. El⁺ represents an electrophile.*

- The electrophile is positively charged or partially positively charged, so it is attracted to the high electron density of the benzene ring.
- A bond forms between the electrophile and the benzene ring.
- This causes the aromatic system to become unstable. One of the carbon atoms becomes positively charged.
- To regain stability, a hydrogen ion is lost.
- The overall reaction is a substitution reaction because the electrophilic reagent has replaced a hydrogen atom in the benzene ring.

Bromination of benzene

Benzene reacts with bromine in the presence of a catalyst of iron(III) bromide (or iron filings and bromine). The overall reaction is:

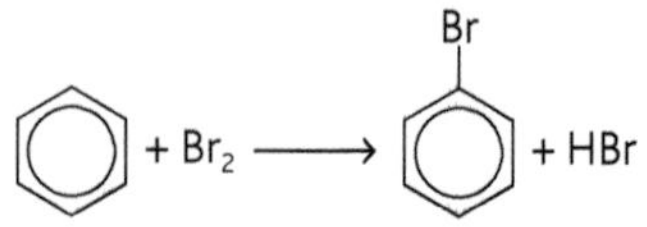

The electrophilic reagent is the positively polarised bromine molecule.

The highly polar iron(III) bromide causes the movement of electrons as shown. We say that the iron(III) bromide is a halogen carrier.

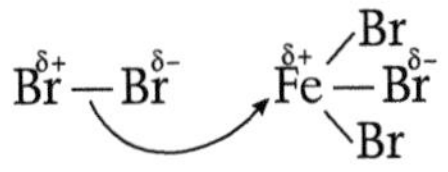

$$Br^{\delta+} \!-\! Br^{\delta-} \qquad Fe^{\delta+}\!-\!Br^{\delta-}$$

The mechanism is shown in Figure 4.1.2.

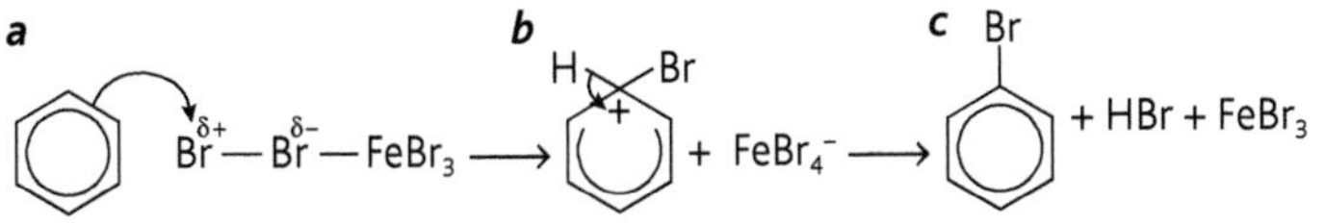

Figure 4.1.2 *The bromination of benzene;* **a** *electrophilic attack;* **b** *intermediate formed;* **c** *loss of H^+ ion and reformation of catalyst*

Did you know?

In the reaction between bromine and benzene, some books show the attacking reagent as Br^+. It is useful to show the bromine with a full charge in this way because it conforms to the general pattern. The reaction shown in Figure 4.1.2 is, however, more likely to occur.

- The electrophile (positively polarised bromine molecule) attacks the delocalised electrons in the benzene ring.
- A bond forms between the bromine and the benzene ring.
- An unstable positively charged intermediate is formed.
- A hydrogen ion is lost and combines with a Br from the $FeBr_4^-$ to form HBr. The $FeBr_3$ catalyst is reformed.

Nitration of benzene

Benzene reacts with a mixture of concentrated nitric and concentrated sulphuric acids. The overall reaction is:

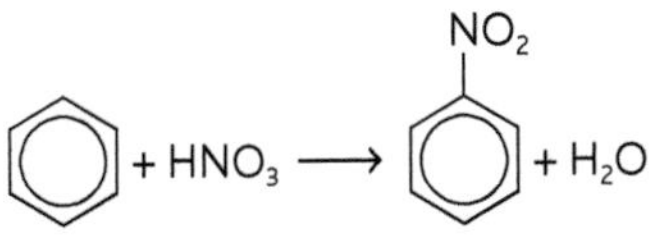

(diagram) $+ HNO_3 \longrightarrow$ (diagram with NO_2) $+ H_2O$

The electrophilic reagent is the nitronium ion, NO_2^+, formed by the nitrating mixture of concentrated nitric and sulphuric acids:

$$HNO_3 + 2H_2SO_4 \rightarrow NO_2^+ + 2HSO_4^- + H_3O^+$$

The mechanism is shown in Figure 4.1.3.

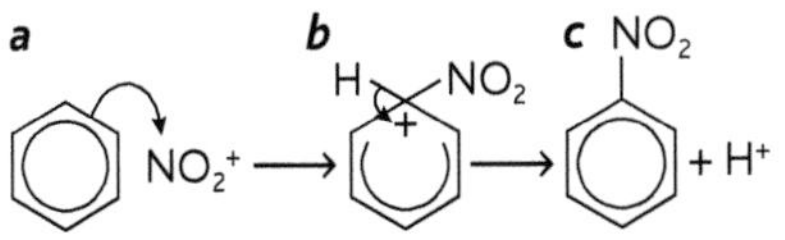

(diagram)

Figure 4.1.3 *The nitration of benzene;* **a** *electrophilic attack;* **b** *intermediate formed;* **c** *loss of H^+ ion*

- The electrophile (the nitronium ion, NO_2^+) attacks the delocalised electrons in the benzene ring.
- A bond forms between the nitro group, NO_2, and the benzene ring.
- An unstable positively charged intermediate is formed.
- A hydrogen ion is lost. This reforms the sulphuric acid ($H^+ + HSO_4^-$).

Key points

- The main mechanism of reaction of arenes involves electrophilic substitution.
- In the electrophilic substitution of arenes, a hydrogen atom in the ring is replaced by Br, NO_2 or another suitable group.
- When benzene is nitrated, the nitronium ion, NO_2^+ is the electrophilic reagent.
- When benzene is brominated, a halogen carrier is required.
- The electrophilic reagent in the bromination of benzene is the polarised bromine molecule.

Substituted arenes

Figure 4.2.1 shows how we name some substituted arenes.

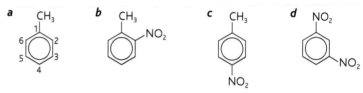

Figure 4.2.1 *Some substituted arenes; a methylbenzene, b methyl-2-nitrobenzene, c methyl-4-nitrobenzene and d 1,3-dinitrobenzene*

The nitration of methylbenzene

Methylbenzene reacts with the same electrophiles as benzene but:

- it is slightly more reactive than benzene
- a mixture of isomers is obtained.

When methylbenzene is nitrated with a mixture of concentrated nitric and sulphuric acids, the isomers produced are:

methyl-2-nitrobenzene 59%

methyl-3-nitrobenzene 4%

methyl-4-nitrobenzene 37%

Because it is mainly the 2- and 4- isomers that are produced, we say that the CH_3 group in methylbenzene is 2-, 4-directing.

This is because the CH_3 group tends to release electrons to the benzene ring. (+ *I* effect – see Section 5.1.) This reduces the positive charge on the intermediate formed and therefore stabilises the intermediate (see Figure 4.2.2).

The possible intermediates formed by reaction at the 2- and 4-positions have the positive charge on the C atom next to the methyl group. The possible intermediates formed at the 3-position have the positive charge on other carbon atoms.

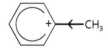

Figure 4.2.2 *A methyl group tends to donate electrons to the benzene ring. The arrow shows the direction of movement of the electrons.*

Bromination and nitration of methylbenzene

In the presence of a halogen carrier, bromine reacts as an electrophilic reagent. A bromine atom is substituted in either the 2- or 4-positions of the benzene ring. A mixture of isomers is obtained. The methyl group is again, 2-, 4-directing.

The nitration and bromination of nitrobenzene

Nitrobenzene reacts with the same electrophiles as benzene but:

■ it is slightly less reactive than benzene (it requires heating to 50°C rather than reaction at room temperature)

■ the 3-isomer is mainly obtained.

Because it is mainly the 3-isomer that is produced, we say that the NO_2 group in nitrobenzene is 3-directing.

This is because the NO_2 group is an electron-attracting group. It tends to withdraw electrons from the benzene ring. ($-I$ effect – see Section 5.1.) This increases the positive charge on the intermediate formed, especially at the 2- and 4-positions. So the intermediate formed at the 3-position is most stable.

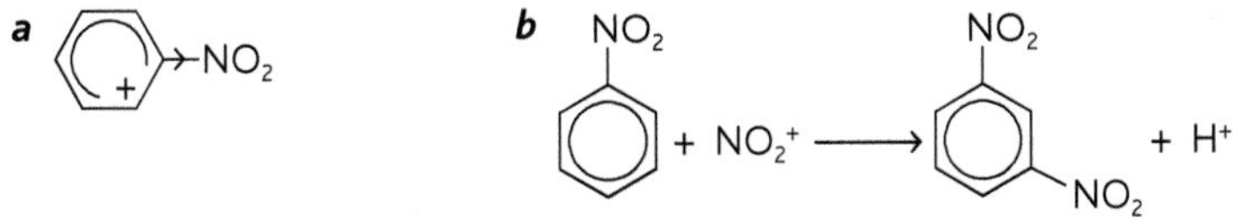

Figure 4.2.3 a *A nitro group tends to withdraw electrons to the benzene ring. The arrow shows the direction of movement of the electrons.* **b** *The equation for the overall reaction of the NO_2^+ ion with nitrobenzene.*

Reaction of nitrobenzene with tin and HCl

Aromatic nitro compounds are reduced by reacting with tin and concentrated hydrochloric acid. Aromatic amines are formed. The actual product is a complex salt of the amine. The amine is formed from this by reaction with alkali. With nitrobenzene, an aromatic amine called phenylamine is formed.

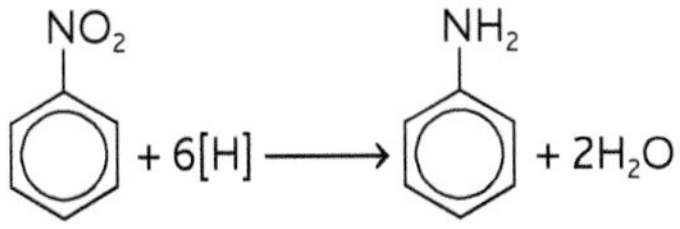

Figure 4.2.4 *A simplified equation for the reduction of nitrobenzene by tin and concentrated hydrochloric acid. [H] represents the reducing power of hydrogen.*

✓ Exam tips

You do **not** have to remember:

1 equations for the reaction of nitrobenzene with tin and hydrochloric acid

2 details of the different intermediates for the nitration and bromination of nitrobenzene and methylnitrobenzene.

Key points

■ When methylbenzene reacts with electrophilic reagents, the products formed are mainly the 2- and 4-isomers of methylnitrobenzene.

■ The methyl group in methylbenzene is 2-, 4-directing because it tends to release electrons to the benzene ring to stabilise the intermediate.

■ Nitrobenzene is reduced to phenylamine by reaction with tin and concentrated hydrochloric acid.

Learning outcomes

On completion of this section, you should be able to:

- describe the reactions of phenols with acid halide (acyl halides), aqueous bromine and sodium hydroxide
- describe the formation of azo compounds and the coupling reaction
- state some uses of azo compounds.

✓ *Exam tips*

Remember that when you write the ring structure of aromatic compounds with substituent groups, you may see the structure written in different ways. For example, two ways of writing phenol are:

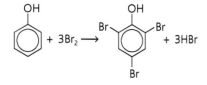

Remember that these are the same structure.

Introduction

Phenol, C_6H_5OH, has an —OH group attached directly to the benzene ring in place of a hydrogen atom. Phenol is only sparingly soluble in water because the large aryl group minimises hydrogen bonding with water molecules. The —OH group does not always react in the same way as the —OH group in alcohols. For example, phenol is acidic, whereas alcohols are not (see Section 5.2).

Reactions of the —OH group in phenols

Reaction with alkalis and reaction with sodium

Alcohols do not react with sodium hydroxide. However, because of its acidic character, phenol does react. It reacts with alkalis to form a salt (called a phenoxide) and water.

For example, with sodium hydroxide, sodium phenoxide is formed:

$$\text{phenol} \quad \text{—OH} + \text{NaOH} \longrightarrow \text{O}^-\text{Na}^+ + \text{H}_2\text{O} \quad \text{sodium phenoxide}$$

The sodium phenoxide is soluble in water because it is ionic.

Phenol reacts with sodium to form sodium phenoxide and hydrogen.

$$2C_6H_5OH + 2Na \rightarrow 2C_6H_5O^-Na^+ + H_2$$

Reaction with acyl halides (acid halides)

Acyl halides (acid halides) are formed when carboxylic acids react with PCl_5 (see Section 3.5). Acyl halides react with phenols to form esters. The OH bond in the phenol is broken. Fumes of HCl are released. Alcohols also react with acyl halides in a similar manner.

$$\underset{\text{ethanoyl chloride}}{CH_3COCl} + \underset{\text{phenol}}{C_6H_5OH} \rightarrow \underset{\text{phenyl ethanoate}}{CH_3COOC_6H_5} + \underset{\text{hydrogen chloride}}{HCl}$$

Substitution reactions in the aromatic ring

Phenol reacts much more rapidly with electrophiles than benzene.

- A lone pair of electrons on the oxygen atom overlaps with the delocalised electrons in the aromatic ring (see Section 5.2).
- This increases the electron density in the aromatic ring.
- So the positive charge on the intermediates is reduced.
- The intermediates are more stable compared with benzene (especially at the 2-, 4- and 6-positions).
- So the reaction occurs under milder conditions and more positions in the ring are substituted.

The reaction of phenol with bromine

Bromine water reacts rapidly with phenol. A white precipitate of 2,4,6-tribromophenol is formed. No halogen carrier is needed.

Diazotisation and coupling reactions

Diazotisation

Phenylamine, $C_6H_5NH_2$, an aromatic amine, reacts with nitrous acid in the presence of hydrochloric acid to form a **diazonium salt** (general formula $RN^+\equiv NX^-$). This process is called **diazotisation**.

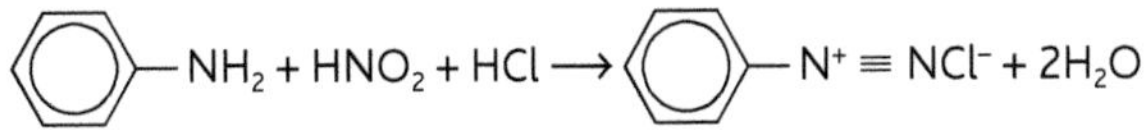

phenylamine   nitrous acid   benzene diazonium chloride

Nitrous acid is unstable and so is made by adding $NaNO_2$ and HCl to the phenylamine. The reaction mixture has to be kept below 10 °C to prevent the diazonium salt decomposing to nitrogen.

Coupling reaction

Benzenediazonium chloride reacts with an alkaline solution of phenol to form an orange dye (4-hydroxyphenylazobenzene). This is called a **coupling reaction**. The temperature of the solution must be kept well below 10 °C to prevent decomposition of the diazonium salt.

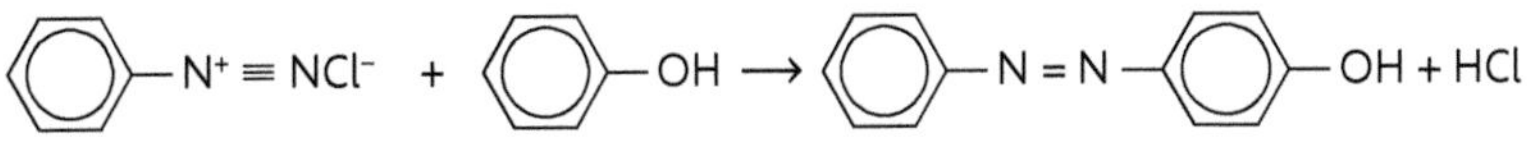

benzene diazonium chloride   phenol   orange dye

Figure 4.3.1 *A coupling reaction is the reaction of a diazonium salt with a phenol under alkaline conditions*

The positive charge on the diazonium ion acts as an electrophile and substitutes into the 4-position of the ring of the phenol.

☑ Exam tips

The important points about the diazotisation and coupling reactions are that:

- aromatic amines and phenols are involved
- HNO_2 is made from $NaNO_2$ and HCl
- the reaction is carried out below 10 °C
- substitution of the diazo compound is at the 4-position in the phenol ring
- azo compound have the N=N link.

Azo dyes

The dye formed in the coupling reaction is an **azo dye**. It contains the —N=N— group attached to two aromatic rings. By using different aromatic amines and different phenols, we can make a variety of brightly coloured dyes. The colour depends on the amine and phenolic compound used.

Did you know?

Diazonium compounds are important as intermediates in organic synthesis. The $N^+\equiv NX^-$ group can be replaced by a variety of other groups, e.g. reaction with KI to produce substituted iodo-compounds, reaction with CN^- to produce aromatic cyanides.

☑ Exam tips

The reaction of phenol with other electrophiles also occurs under milder conditions and with more positions in the ring being substituted. For example, nitration only requires dilute nitric acid (no sulphuric acid is needed) and 2,4,6-trinitrophenol is formed.

Key points

- Phenols have one or more —OH groups attached directly to the benzene ring.
- The —OH group in phenol reacts with NaOH to form a salt and with acid halides to form esters.
- The electrophilic substitution in the ring of phenol is rapid due to activation of the ring by the —OH group.
- Phenylamine reacts with nitrous acid below 10 °C to form benzenediazonium chloride.
- A coupling reaction occurs when diazonium salts react with phenols to form azo dyes.

Revision questions

Answers to all revision questions can be found on the accompanying CD.

1 A compound has the following properties:
   i reacts readily with bromine water
   ii turns blue litmus paper pink
   iii does not cause effervescence when mixed with aqueous sodium carbonate.

   Which of the following compounds most accurately exhibits all of the above properties?
   A C_6H_5OH
   B $C_6H_5NH_2$
   C C_6H_5COOH
   D $C_6H_5CONH_2$

2 Which of these statements best explains the weak acidity of phenol?
   A The $C_6H_5O^-$ ion is a weak conjugate base.
   B The $C_6H_5O^-$ ion is stabilised via delocalisation of charge into the benzene ring.
   C C_6H_5OH is a stronger acid than an aqueous solution of carbon dioxide (carbonic acid).
   D C_6H_5OH will effervesce with potassium carbonate solution.

3 Which of the following reagents best differentiates between propan-1-ol and propan-2-ol?
   A $Cr_2O_7{}^{2-}/H^+(aq)$
   B $NaOH(aq)/I_2$
   C conc. H_2SO_4
   D $Na(s)$

4 Which of the following classes of compounds is produced by the dehydration of alcohols?
   A alkanes
   B alkynes
   C alkenes
   D arenes

5 Which compound is most likely to be formed when phenol is reacted with bromine in an organic solvent?
   A 4-bromophenol
   B 2,4,6-tribromophenol
   C 3-bromophenol
   D 2,3-dibromophenol

6 What is the correct order of increasing acidity of the compounds below?

| X | Y | Z |

   A Z Y X
   B Z X Y
   C Y Z X
   D X Y Z

7 A compound **A**, C_3H_8O, when refluxed with acidified potassium dichromate(vi) produced a liquid **B**, with a sharp smell, which effervesced with sodium carbonate solution. Another compound **C**, containing two carbon atoms, when heated with concentrated sulphuric acid at temperatures of about 160 °C produced a sweet-smelling gas **D**, which decolourised aqueous bromine. **C** also produced yellow crystals when gently warmed with sodium hydroxide and iodine. **C** when refluxed with **B** in the presence of concentrated sulphuric acid produced a sweet-smelling compound. **E** is used in the perfumery industry.
   a Explain the arguments used to deduce the names and structural formulae of the compounds **A**, **B**, **C**, **D** and **E**.
   b Give the equations for the reactions of **C** and **B**, and also **B** and sodium carbonate.
   c Write the type of reaction which produces **D** and **E**.

8 a Explain the terms 'primary', 'secondary' and 'tertiary' as applied to alcohols.
   b There are four structural isomeric alcohols with formula $C_4H_{10}O$.
      i Write the names and structural formulae of these alcohols and classify them according to terms in **a** above.
      ii Describe laboratory tests to distinguish between these groups of compounds.
   c One of these isomers when heated with an excess of concentrated sulphuric acid produces three different compounds.
      i State the type of reaction involved.
      ii Explain the production of these compounds.

9 a Explain the characteristic difference in effect when the hydroxyl group, OH, is bonded to an alkyl and the benzene ring respectively.

 b Explain, with the aid of an equation, what you would see when phenol is added to:
  i ethanoyl chloride
  ii an aqueous solution of bromine
  and write the structural formula for the product.

 c A sequence of two reactions is shown below:

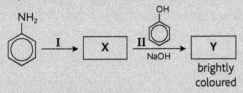

  i Write the names and structural formulae for the compounds **X** and **Y**.
  ii Write the reagents and conditions for reaction **I**.
  iii Name the type of reactions in **I** and **II**.

10 Compare and contrast the reactions of propan-1-ol and phenol, with the following reagents, giving equations whenever possible and describing appropriate observations:
  i water
  ii blue litmus paper
  iii potassium hydroxide
  iv phosphorus(v) chloride
  v ethanoic acid.

5 Organic acids and bases

5.1 Carboxylic acids and acidity

Learning outcomes

On completion of this section, you should be able to:

- explain the difference in acidity of alcohols and carboxylic acids

- understand why increasing the substitution of Cl onto the carbon atom next to the —COOH group results in stronger acidity

- understand the effect of the conjugative (mesomeric) effect and inductive effect on determining the acidity of various substituted carboxylic acids and in alcohols.

Did you know?

Trichloroethanoic acid is used to remove skin imperfections and wrinkles. The cosmetic treatment, known as a chemical peel, must be carried out under medical supervision because this acid is corrosive. After treatment the patient appears to have sunburn!

K_a values of some carboxylic acids

Carboxylic acids are weak acids. They are not completely ionised (dissociated) in solution. The position of equilibrium is well over to the left. For example, ethanoic acid (CH_3COOH).

$$CH_3COOH + H_2O \rightleftharpoons CH_3COO^- + H_3O^+$$

The table shows some K_a and pK_a values for different carboxylic acids.

Carboxylic acid	K_a mol dm^{-3}	pK_a
ethanoic acid, CH_3COOH	1.7×10^{-5}	4.77
propanoic acid, CH_3CH_2COOH	1.3×10^{-5}	4.89
chloroethanoic acid, $CH_2ClCOOH$	1.3×10^{-3}	2.89
dichloroethanoic acid, $CHCl_2COOH$	5.0×10^{-2}	1.30
trichloroethanoic acid, CCl_3COOH	2.3×10^{-1}	0.64

The values of K_a and pK_a give us information about the position of equilibrium (see *Unit 1 Study Guide*, Section 9.3). The higher the value of K_a and the lower the value of pK_a:

- the further the position of equilibrium is to the right (in favour of products)
- the greater the acidity of the carboxylic acid.

We can explain these differences by referring to two effects:

- The conjugative effect (also known as the mesomeric or resonance effect).
- The inductive effect.

The conjugative effect

The **conjugative effect** occurs in molecules with multiple bonds, usually when the structure at first sight appears to be of alternating double and single bonds e.g. benzene. The π orbitals overlap and form a delocalised system over three or more atom centres (see Section 1.1). The effect makes the bond lengths and bond strengths intermediate between ordinary single bonds and ordinary double bonds. The effect also makes the structure more stable. Other compounds showing this effect are:

- the carbon–oxygen bonds in carbonate ions
- the carbon–oxygen bonds in carboxylate ions derived from carboxylic acids.

Taking carboxylic acids as an example:

- Oxygen atoms are more electronegative than carbon atoms or hydrogen atoms.
- The bonding pairs of electrons tend to move towards the atoms as shown by the arrows in the diagram on page 50.
- The O—H bond is weakened so that it is possible for a proton to be lost.
- The carboxylate ion formed exists as a resonance hybrid of two extreme forms. The actual form is somewhere in between these two extremes (Figure 5.1.1).
- The carboxylate ion is relatively stable, so the ability of the H⁺ ion to combine with it is reduced.

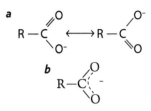

Figure 5.1.1 *Conjugative effect in the carboxylate ion; a The extreme resonance forms; b The resonance hybrid*

The inductive effect

Bonds between unlike atoms are polarised due to the difference in electronegativity between the atoms (see *Unit 1 Study Guide*, Section 2.5). The polarisation is represented by an arrow pointing towards the more electronegative atom. Groups of atoms can also exert an electron-attracting or withdrawing effect on the electrons around a particular atom such as a carbon atom. This is called the **inductive effect**.

- Atoms or groups more electronegative than carbon withdraw electrons from around the carbon atom. This is called the negative inductive effect (– *I* effect). For example chlorine bonded to carbon.

$$Cl{\leftarrow}\overset{|}{\underset{|}{C}}{-}$$

- Atoms or groups less electronegative than carbon tend to donate electrons to the carbon atom. This is called the positive inductive effect (+*I* effect). For example alkyl groups bonded to carbon.

$$CH_3{\rightarrow}\overset{H}{\underset{H}{C}}{-}H \quad CH_3{\rightarrow}\overset{CH_3}{\underset{H}{C}}{-} \quad CH_3{\rightarrow}\overset{CH_3}{\underset{CH_3}{C}}{-}$$

→ increasing +*I* effect

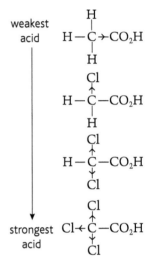

Figure 5.1.2 *As the number of electron-withdrawing groups on the COOH carbon atom increases, the strength of the acid increases*

Comparing acidity: chloroethanoic acids

The table opposite shows the pK_a values of chloroethanoic acid, dichloroethanoic acid and trichloroethanoic acid. We can explain the difference in acidity of the chloroethanoic acids by the inductive and conjugative effects.

- The more Cl atoms that are substituted in the CH$_3$ group in ethanoic acid, the greater is the electron-withdrawing (– *I*) effect on the C atom of the —COOH group.
- The greater the –*I* effect, the more the electrons are withdrawn from the carbon and the more the electrons in the O—H bond are drawn towards the oxygen atom.
- The more the electrons are drawn towards the O atom in the OH bond, the weaker the bond and the more likely it is that a H⁺ ion will be formed.
- The greater the –*I* effect, the greater is the delocalisation of the negative charge.
- So the greater is the conjugative effect and the more likely it is that a H⁺ ion will be formed.

Key points

- The —COO⁻ ion in carboxylic acids is stabilised by the conjugative (mesomeric) effect.
- The acid strength of carboxylic acids increases if the C next to the COOH group has electron-withdrawing atoms bonded to it.
- Carboxylic acids are stronger acids than alcohols because of the effect of the conjugative and inductive effects on the —COOH group.
- The greater the number of Cl atoms substituted on the carbon atom next to the —COOH group, the greater the acidity of the carboxylic acid.

Introduction

Ethanol and ethanoic acid are soluble in water. This is because they can form extensive hydrogen bonding with water. Phenol is only sparingly soluble in water and is a solid at room temperature. Its higher molar mass and large aryl ring reduce its hydrogen bonding capacity with water. Phenol, however, although less acidic than ethanoic acid is more acidic than ethanol. Why is this?

The acidity of compounds with —OH groups

The table shows the pK_a values for some compounds containing the —OH group. The lower the value of pK_a, the more acidic the compound.

Compound	Formula	pK_a
ethanoic acid	CH_3COOH	4.8
phenol	C_6H_5OH	10.0
ethanol	C_2H_5OH	16.0

Phenol is a very weak acid. It ionises slightly in water to form the phenoxide ion.

$$C_6H_5OH(s) + aq \rightleftharpoons C_6H_5O^-(aq) + H^+(aq)$$
$$\text{phenol} \qquad \text{phenoxide ion}$$

Phenol is a much weaker acid than ethanoic acid but it is more acidic than water or ethanol.

- An aqueous solution of phenol has a pH just below 7 whereas alcohol and water appear neutral with narrow range universal indicator paper.
- Phenol is too weak an acid to liberate carbon dioxide from carbonates whereas ethanoic acid will liberate carbon dioxide.
- Neither ethanol nor water will neutralise sodium hydroxide but phenol does.

Ethoxide, phenoxide and ethanoate ions

We can explain the differences in acidity of ethanol, phenol and ethanoic acid by comparing the structure of the phenoxide, ethoxide and ethanoate ions.

Ethanol

- The charge on any ethanoate ions formed is concentrated on the oxygen atom because of the positive inductive effect of the ethyl group and the electronegativity of the O atom.
- This increases the negative charge on the oxygen atom. So any ethoxide ion formed is more likely to accept a H⁺ ion.
- There is no conjugative effect to stabilise the ethanoate ion since ethanol does not have a $C=O$ group.
- The position of equilibrium is so far over to the left that the ethanol exists as undissociated molecules with hardly any ethanoate ions.

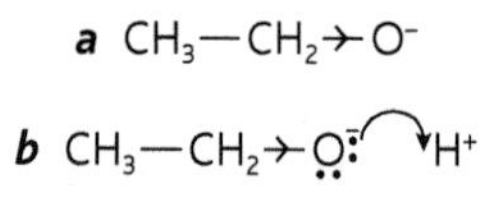

a $CH_3-CH_2 \rightarrow O^-$

b $CH_3-CH_2 \rightarrow \ddot{O}: \curvearrowright H^+$

Figure 5.2.1 a The ethoxide ion has negative charge concentrated on the oxygen; **b** A hydrogen ion is readily attracted to the highly charged negative ion

Phenol

- In the phenoxide ion, one of the lone pairs of electrons in a p-orbital on the oxygen atom overlaps with the delocalised electrons in the phenol ring.

- An extended delocalised system is formed which includes the oxygen (see Figure 5.2.2).

- The conjugative (resonance) effect is increased because the delocalised electrons can move over a larger area.

- The extra delocalisation reduces the electron density on the oxygen. The charge is spread over the whole ion rather than being confined to the oxygen.

- The increased conjugative effect stabilises the phenoxide ion.

- The conjugative effect is greater than the inductive effect, so the movement of the electrons in the C—O bond is towards the phenol ring rather than towards the oxygen.

- The position of equilibrium is further to the right compared with ethanol.

- H^+ ions are not as strongly attracted to the phenoxide ion. So the phenoxide ion is less likely to form undissociated molecules than is the ethoxide ion.

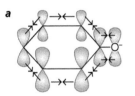

Figure 5.2.2 The extended delocalised ring system in phenol; **a** The isolated p-orbitals; **b** The delocalised ring system; **c** The movement of electrons in the C—O bond is towards the ring

Ethanoic acid

- Some of the p-electrons in the COO^- group are delocalised.

- This delocalisation reduces the electron density on the oxygen (the charge is spread over the whole ion rather than being confined to the oxygen of the O—H group).

- The conjugative effect stabilises the ethanoate ion.

- The conjugative effect is greater than in phenol, so the stabilisation of the ion is greater.

- The position of equilibrium is further to the right compared with phenol.

- H^+ ions are not as strongly attracted to the ethanoate ion. So the ethanoate ion is less likely to form undissociated molecules than is the phenoxide ion.

Figure 5.2.3 The delocalised electrons in the COO^- ion; **a** The isolated p-orbitals; **b** The delocalised electrons in the carboxylate ion

Key points

- Phenol is weakly acidic but more acidic than ethanol or water.

- Phenol is less acidic than carboxylic acids.

- The acidity in phenols is due to the delocalisation of charge on the phenoxide ion with the delocalised electrons in the phenol ring.

- The alkoxide ion does not have delocalised charge associated with the O atom. So the O^- readily accepts a hydrogen ion to form a molecule.

5.3 Amines, amides and acyl halides

Learning outcomes

On completion of this section, you should be able to:

- state the difference in basic character of **aliphatic** amines, aromatic amines and amides
- explain the difference in basic character of aliphatic amines, aromatic amines and amides in terms of the inductive and conjugative effects.

Introduction

Amines are compounds with the —NH_2 functional group. Amines can be thought of as ammonia molecules in which one or more of the hydrogen atoms has been substituted by an alkyl group. **Amides** have the —$CONH_2$ functional group. The structures of some of these compounds are shown below.

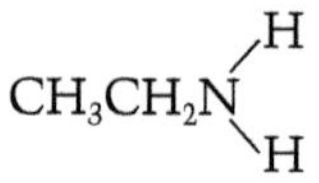

ethylamine
(a primary amine)

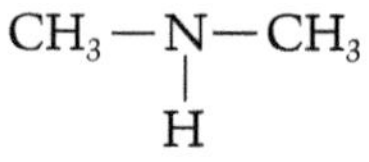

dimethylamine
(a secondary amine)

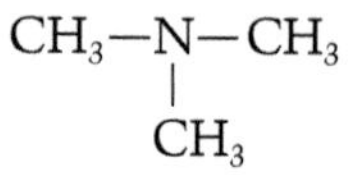

trimethylamine
(a tertiary amine)

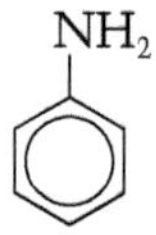

phenylamine
(an aromatic amine)

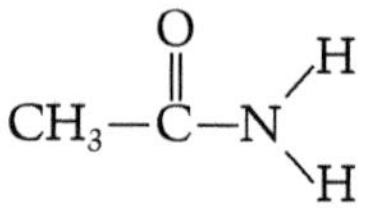

ethanamide
(an amide)

Basic character of amines

Amines are weak bases. The position of equilibrium lies to the left. For example, methylamine (CH_3NH_2).

$$CH_3NH_2 + H_2O \rightleftharpoons CH_3NH_3^+ + OH^-$$

Amines react with acids to form salts.

$$CH_3NH_2 + HCl \rightarrow CH_3NH_3^+Cl^-$$

The table shows the pK_b values for ammonia and some amines. The lower the value of pK_b, the more basic is the compound.

Compound	Formula	pK_b
Ammonia	NH_3	4.7
Methylamine	CH_3NH_2	3.4
Dimethylamine	$(CH_3)_2NH$	3.3
Phenylamine	$C_6H_5NH_2$	9.4

The strength of a base depends on the availability of the lone pairs of electrons on the nitrogen atom to bond to a H^+ ion.

Methylamine is a stronger base than ammonia because:

- the methyl group is electron-donating (+*I* effect). It releases electrons to the nitrogen atom.
- so the lone pair on the N atom of the amine has a higher electron density compared with the electron density of the N atom in ammonia.
- the lone pair on the N atom of methylamine is better at accepting a H^+ ion from water (compared with the N atom of ammonia).

Exam tips

Remember that the values of pK_b gives us information about the position of equilibrium. The lower the value of pK_b:

- the further the position of equilibrium is to the right (in favour of products)
- the greater the basicity of the amine.

Secondary amines have two alkyl groups so are stronger bases than the corresponding primary amines. Tertiary amines, however, are weaker bases than ammonia.

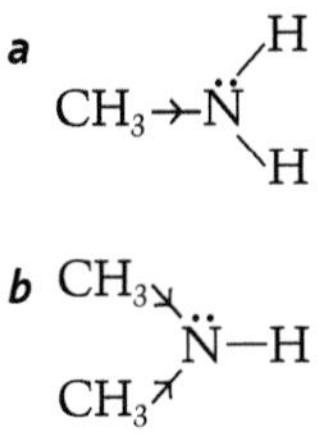

Why is phenylamine a very weak base?

- Phenylamine is a very weak base compared with ammonia and alkyl amines. This is because of the conjugative (resonance effect).
- In the phenylamine molecule the lone pair of electrons in a p-orbital on the nitrogen atom overlaps with the delocalised electrons in the ring.
- An extended delocalised system is formed which includes the nitrogen atom (see Figure 5.3.2).
- The conjugative (resonance) effect is increased because the delocalised electrons can move over a larger area.
- The extra delocalisation reduces the electron density on the nitrogen atom.
- The increased conjugative effect stabilises the phenylamine molecule.
- The conjugative effect is greater than the inductive effect, so the movement of the electrons in the C—N bond is towards the ring rather than towards the nitrogen atom.
- The position of equilibrium is much further to the left compared with ammonia.
- So H^+ ions from water are not as strongly attracted to the nitrogen atom compared with the stronger attraction to the N atom in ammonia.

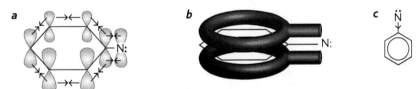

Figure 5.3.2 The extended delocalised ring system in phenylamine; **a** The isolated p-orbitals; **b** The delocalised ring sytem; **c** The movement of electrons in the C—N bond is towards the ring.

Amides

Amides are much weaker bases than amines. For example the pK_b of ethanamide is 14.1. Amides are neutral to litmus and do not react with hydrochloric acid to form salts. This is because:

- a p-orbital on the nitrogen atom interacts with a p-orbital on the adjacent carbon atom
- there is considerable delocalisation between the O, C and N atoms which does not occur in ammonia or alkyl amines.

This reduces the ability of the lone pair on the nitrogen atom to accept a H^+ ion from water.

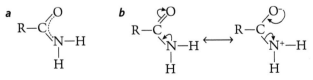

Figure 5.3.3 a The conjugative (resonance) effect provided by the carbonyl group in amides stabilises the structure; **b** The resonance hybrids of an amide

Figure 5.3.1 a Methylamine is a stronger base than ammonia because of the $+I$ effect of the methyl group. **b** Dimethylamine is a stronger base than methylamine because it has two $+I$ effects.

✓ Exam tips

Phenylamine is the starting point for making diazonium compounds and azo dyes. Make sure that you know these reactions by referring back to Section 4.3.

Key points

- Amines behave as bases because the lone pair of electrons on the N atom tends to accept a proton.

- Amines are stronger bases than ammonia because the electron-releasing alkyl group increases the electron density on the nitrogen atom.

- Phenylamine is a weaker base than ammonia because the lone pair of electrons on the nitrogen atom in phenylamine is delocalised with the delocalised electrons in the aromatic ring.

- Amides are neutral to litmus and do not react with acids.

Learning outcomes

On completion of this section, you should be able to:

■ explain the acid–base properties of amino acids

■ understand the term 'zwitterions'

■ state that amino acids are naturally occurring compounds which make up proteins.

The structure of amino acids

Amino acids contain the amino group ($—NH_2$) and the carboxylic acid group ($—COOH$). Amino acids will undergo most of the reactions of amines and carboxylic acids.

The amino acids most often found in living organisms have the $—NH_2$ bonded to the carbon atom next to the $—COOH$ group (α-amino acids or 2-amino acids). Most amino acids exist as optical isomers. Figure 5.4.1 shows the general structure of an amino acid. The R group can vary greatly; it can be acidic, basic or neutral. Neutral R groups can be polar or non-polar (see Figure 5.4.2). You do not have to know details of these types of amino acid.

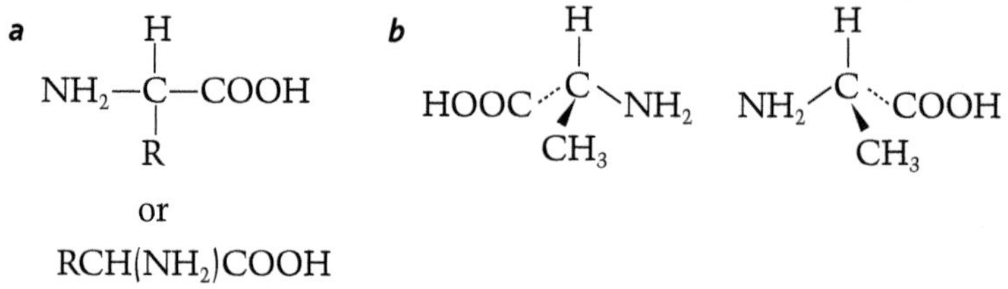

Figure 5.4.1 a General structure of an amino acid; **b** The two optical isomers of the amino acid, alanine

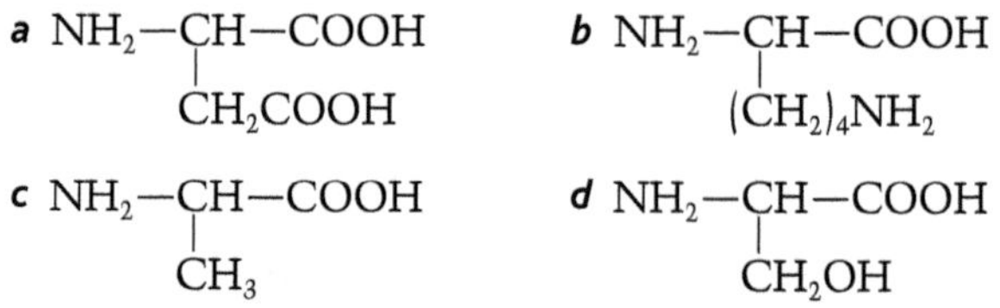

Figure 5.4.2 Amino acids with **a** an acidic side chain; **b** a basic side chain; **c** a neutral non-polar side chain and **d** a neutral polar side chain

Zwitterions

The basic $—NH_2$ and acidic $—COOH$ group in amino acids can react with each other. An ion carrying two charges, one positive and the other negative is formed. This type of ion is called a **zwitterion**. A zwitterion is an electrically neutral species with a positive and negative ion in two different parts of the species. The positive and negative charges cancel and hence the 'molecule' is neutral.

In the formation of a zwitterion in amino acids the $—COOH$ group has lost a H^+ ion and the $—NH_2$ group has gained a H^+ ion.

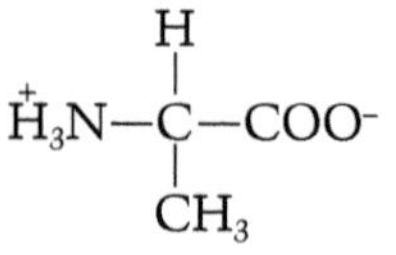

Figure 5.4.3 The zwitterion of the amino acid alanine

Amino acids are crystalline solids because of the relatively strong ionic forces of attraction between the positive and negative parts of adjoining 'zwitterions'.

Did you know?

The word *zwitterion* comes from the German word 'zwitter' which means hybrid. Substances containing zwitterions are sometimes called ampholytes or amphoteric electrolytes. A number of chemicals from plants such as alkaloids form zwitterions.

Acid–base properties of amino acids

In solution, amino acids show both acidic and basic properties.

- The —COOH group reacts with alkalis to form a metal salt.
- The —NH$_2$ group reacts with acids to form salts.

Amino acids act as buffer solutions because they have a supply of weak acid and weak base within the same molecule (see *Unit 1 Study Guide*, Section 9.7).

The charge on the amino acid depends on the type of solution in which it is placed. The following argument applies for amino acids with a neutral side chain.

In solution in water (neutral solution)

- Both the —NH$_2$ and —COOH groups are ionised:
  $$+NH_3—CH(R)—COO^-$$
- The positive charge on the $^+NH_3$ ion balances the negative charge on the COO$^-$ ion.
- The solution is electrically neutral.

In acidic solution

- The acid provides hydrogen ions, e.g. from hydrochloric acid.
- The lone pair on the N atom of the —NH$_3^+$ group is already protonated in the zwitterion. The addition of H$^+$ keeps this group positively charged.
- The —COO$^-$ ion is a proton acceptor.
- The amino acid is positively charged.

$$H_3\overset{+}{N}—\underset{R}{\overset{H}{C}}—COO^- + H^+ \longrightarrow H_3\overset{+}{N}—\underset{R}{\overset{H}{C}}—COOH$$

In alkaline solution

- The alkali provides OH$^-$ ions, e.g. from sodium hydroxide.
- The —NH$_3^+$ group in the zwitterion acts as a proton donor to the hydroxide ion since it is positively charged.
- The —COO$^-$ group remains negatively charged.
- The amino acid is negatively charged.

$$H_3\overset{+}{N}—\underset{R}{\overset{H}{C}}—COO^- + OH^- \longrightarrow H_2N—\underset{R}{\overset{H}{C}}—COO^- \; (+H_2O)$$

Key points

- The amino acids which go to make up proteins have —COOH and —NH$_2$ groups which are attached to the same carbon atom.
- The general formula for an amino acid is NH$_2$CH(R)COOH.
- The R group in amino acids can be acidic, basic or neutral.
- The —NH$_2$ group of an amino acid interacts with the —COOH group to form a zwitterion, one end of which is positively charged and the other negatively charged.

Did you know?

Amino acids may have acidic or basic side chains. When dissolved in water, an amino acid with an acidic side chain behaves as a weak acid. This is because there are two acidic —COOH groups and only one basic —NH$_2$ group.

$$NH_2—\underset{CH_2COOH}{\overset{H}{C}}—COOH$$

When dissolved in water, an amino acid with a basic side chain behaves as a weak base. This is because there are two basic NH$_2$ groups and only one acidic —COOH group.

$$NH_2—\underset{(CH_2)_4NH_2}{\overset{H}{C}}—COOH$$

6 Polymers

6.1 Addition polymerisation

Learning outcomes

On completion of this section, you should be able to:

- describe the characteristics of addition polymerisation
- describe the formation of poly(ethene), polyvinyl chloride and poly(tetrafluoroethene) from their monomers.

Did you know?

Plastics are examples of polymers. The word plastic is derived from the Greek word 'plastikos' which means 'that which can be formed'. In other words, 'that which can be moulded'. In some desert areas, plastic 'trees' have been planted to help trap moisture and improve the growth of real trees planted between them.

Introduction

Polymers are very large molecules built up from a large number of small molecules called **monomers**. The process of joining monomers together to form polymers is called **polymerisation**. There are two types of polymerisation:

- addition polymerisation
- condensation polymerisation.

Addition polymerisation

In addition polymerisation:

- the monomers join together by addition reactions (usually involving free radicals)
- the monomers are usually unsaturated carbon compounds containing the C=C group
- the polymer is the only product of the reaction.

Some examples of addition polymerisation

Poly(ethene)

- The monomers are ethene, $CH_2{=}CH_2$.
- The polymer is called poly(ethene). The common name is polythene.
- The π-bond in each monomer breaks and bonds with the next ethene molecule to form a chain many thousands of carbon atoms long.
- The conditions needed for the reaction depends whether the polymer required is of low density or high density:
  - low-density poly(ethene) for plastic bags and insulation for electric cables: high pressure and high temperature are required
  - high-density poly(ethene) for buckets and bottles: lower temperature and pressures with a special catalyst.

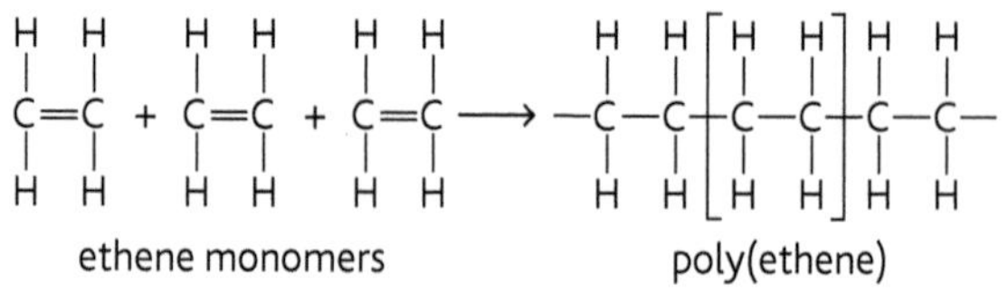

Figure 6.1.1 The formation of part of a poly(ethene) chain from three ethene monomers. The square brackets show the repeating unit.

You will notice that the polymer consists of repeating units derived from the monomer. The **repeating unit** in a polymer is the smallest group of atoms derived from the monomer which when joined gives the structure of the polymer. Note that the repeating unit for poly(ethene) is based on ethene rather than the simplest repeating unit (CH_2).

Poly(tetrafluoroethene) (PTFE)

- This polymer is used as a non-stick coating for saucepans. The most common commercial brand of PTFE is 'Teflon®'. The monomer is tetrafluoroethene, $CF_2 = CF_2$.

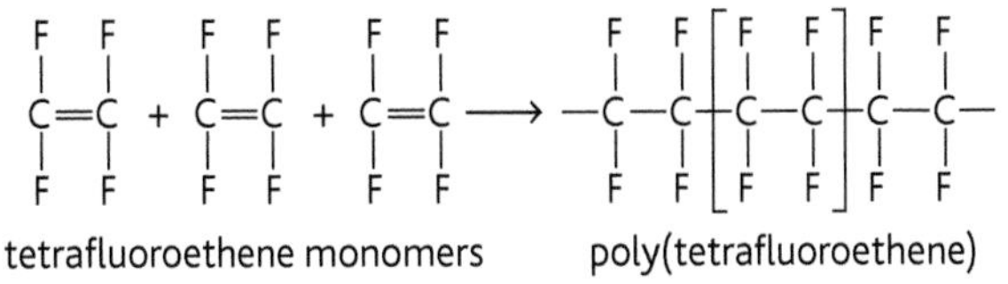

tetrafluoroethene monomers   poly(tetrafluoroethene)

Figure 6.1.2 *The formation of part of a poly(fluoroethene) chain from three fluoroethene monomers. The square brackets show the repeating unit.*

Polyvinyl chloride

Polyvinyl chloride is the common name for poly(chloroethene). The monomer is chloroethene, $CHCl = CH_2$. Figure 6.1.3 shows a shorthand way of writing the polymer chain:

- an 'n' is placed in front of the monomer to show that there are a large number of them to be joined
- only the repeating unit on the polymer is shown
- an 'n' is placed at the bottom right of the repeating unit to show that it is repeated many times.

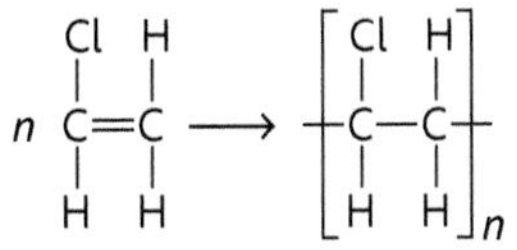

Figure 6.1.3 *A shorthand way of showing the formation of part of a poly(chloroethene) chain from a large number 'n' of monomers*

Some more examples of polymers

Monomer	Repeating unit	Polymer name	Common name
propene, $CH_3CH = CH_2$	$\begin{bmatrix} CH_3 \\ \mid \\ CH-CH_2 \end{bmatrix}_n$	poly(propene)	polypropylene
phenylethene, $C_6H_5CH = CH_2$	$\begin{bmatrix} C_6H_5 \\ \mid \\ CH-CH_2 \end{bmatrix}_n$	poly(phenylethene)	polystyrene
propenenitrile, $CNCH = CH_2$	$\begin{bmatrix} CN \\ \mid \\ CH-CH_2 \end{bmatrix}_n$	poly(propenenitrile)	polyacrylonitrile

Key points

- Polymers are very large molecules built up from a large number of small molecules called monomers.

- Addition polymerisation is when unsaturated monomers bond to each other by addition reactions.

- The repeating unit in a polymer is the smallest group of atoms derived from the monomer which when joined gives the structure of the polymer.

Learning outcomes

On completion of this section, you should be able to:

- describe the characteristics of condensation polymerisation
- describe the formation of Terylene and nylon-6,6 from their monomers.

Condensation reactions

A condensation reaction occurs when two molecules react and a small molecule, such as H_2O or HCl, is eliminated (given off). The formation of an ester from a carboxylic acid and an alcohol or from an acid chloride and an alcohol are examples of condensation reactions. For example:

$$CH_3COOH + CH_3OH \rightleftharpoons CH_3COOCH_3 + H_2O$$
ethanoic acid   methanol   methyl ethanoate   water

$$CH_3COCl + CH_3OH \rightleftharpoons CH_3COOCH_3 + HCl$$
ethanoyl chloride   methanol   methyl ethanoate   hydrogen chloride

Polyesters

Polyesters are polymers with many ester linkages, —COO—. In order to make a polyester, we need to combine:

- a carboxylic acid with at least two —COOH groups (a dicarboxylic acid) with
- an alcohol with at least two —OH groups (a diol).

✅ Exam tips

When writing formulae for polyesters, make sure that the C=O of the COOH group is the correct way round. It should be next to the carbon–hydrogen 'backbone' derived from the carboxylic acid, e.g.

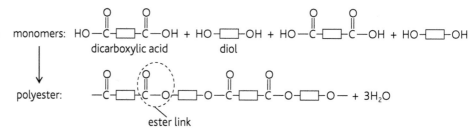

Figure 6.2.1 *Making a polyester from a diol and a dicarboxylic acid. The boxes represent the carbon–hydrogen 'backbone' in each molecule, e.g. —$CH_2CH_2CH_2$—*

Terylene

Terylene is made from benzene-1,4-dicarboxylic acid and ethane-1,2-diol.

The conditions required are catalyst of antimony(III) oxide and a temperature of 280 °C.

Figure 6.2.2 *Making Terylene; **a** the monomers; **b** part of the polymer chain of Terylene. The repeating unit is shown in brackets.*

Polyamides

An amide is formed when a carboxylic acid or acid chloride reacts with an amine. For example:

$$CH_3COCl + CH_3NH_2 \rightleftharpoons CH_3CONHCH_3 + HCl$$
ethanoyl chloride   methylamine   *N*-methyl   hydrogen chloride
ethanamide

Polyamides are polymers with many amide linkages, —CONH—. In order to make a polyamide, we need to combine:

■ a carboxylic acid with at least two —COOH groups (a dicarboxylic acid) with

■ an amine with at least two —NH$_2$ groups (a diamine).

Did you know?

Kevlar is a polyamide made by reacting benzene-1,4-dicarboxylic acid with benzene-1,4-diamine. For its mass, it is five times stronger than steel and is fire resistant. It is used for protective clothing such as bullet-proof vests and fire-resistant clothing.

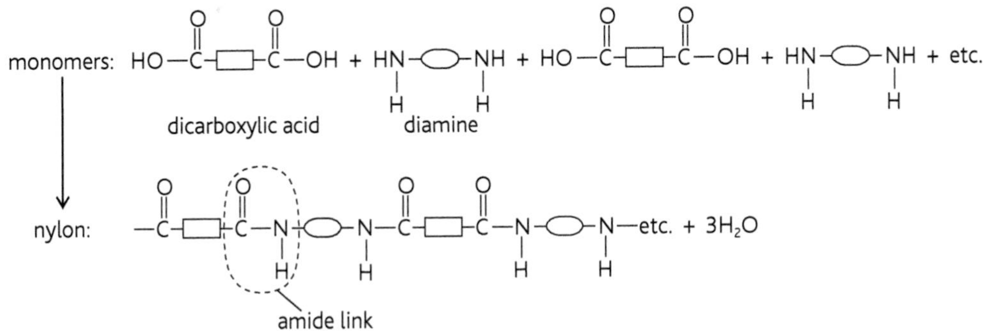

Figure 6.2.3 *Making a polyamide from a dicarboxylic acid and a diamide. The boxes represent the carbon–hydrogen 'backbone' in each molecule, e.g. —CH$_2$CH$_2$CH$_2$—*

Nylon-6,6

Nylon-6,6 is made by reacting hexanedioic acid with the diamide, 1,6-diaminohexane.

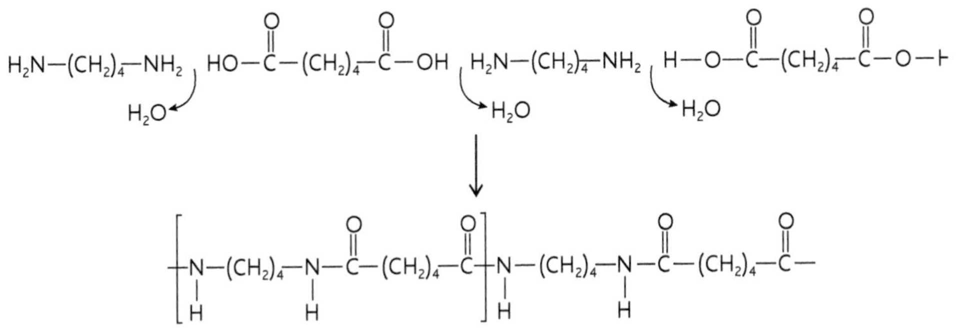

Figure 6.2.4 *Making nylon-6,6; **a** The monomers; **b** Part of the polymer chain of nylon. The repeating unit is shown in brackets.*

The '6,6' in the name of nylon refers to the number of carbon atoms in each monomer unit. Different types of nylon can be made from different monomers.

In the school laboratory, we can use hexanedioyl dichloride $ClOC(CH_4)_2COCl$ in place of hexanedioic acid because the reaction is faster. But this method is too expensive to be used for mass production of nylon.

Key points

■ Condensation polymerisation involves loss of a small molecule, e.g. H$_2$O or HCl, when two types of monomer react.

■ Polyesters, e.g. Terylene and polyamides, e.g. nylon are formed by condensation polymerisation.

■ Polyesters can be made from dicarboxylic acids and diols.

■ Polyamides can be made from dicarboxylic acids and diamines.

Learning outcomes

On completion of this section, you should be able to:

- draw a polymer from a given monomer
- deduce a monomer or monomers from a given polymer.

✅ Exam tips

Note that there are $(2n-1)$ water molecules eliminated in condensation polymerisation because (i) there are two monomers $(2n)$ and (ii) there is one bond fewer than the number of monomers which combine, e.g. for every eight monomers combining there are only seven bonds.

Simplifying structures

We can simplify the way of writing polymerisation reactions by:

- writing an 'n' in front of each monomer to represent a large number
- showing the repeating unit in the polymer in square brackets, followed by 'n' at the bottom right-hand corner
- drawing continuation bonds in the polymer.

Example 1: Addition polymerisation of ethane:

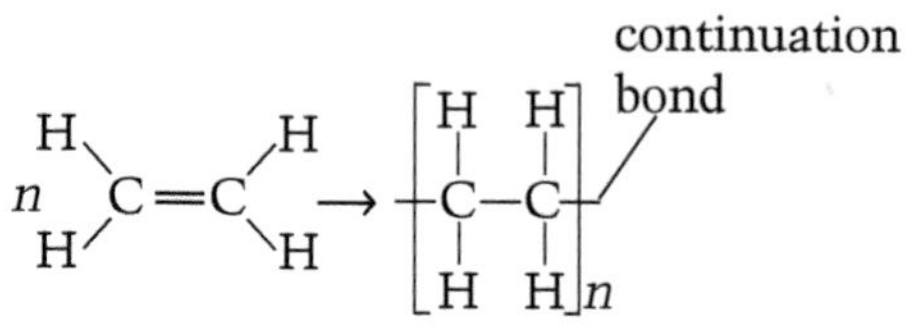

Example 2: Condensation polymerisation to make Terylene:

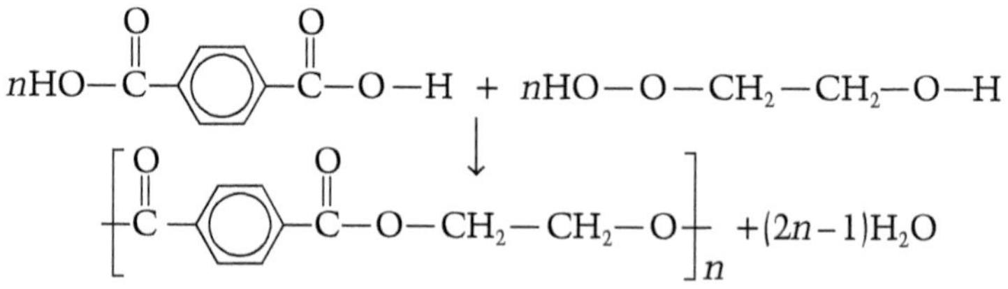

From monomer to polymer

Addition polymers, e.g. poly(propene) from propene

To draw the structure of the polymer:

- Rearrange the structure if necessary to make the atoms stick out vertically from the $C{=}C$ bond (see Figure 6.3.1).
- Draw the structure of the monomer but change the double bond to a single bond.
- Put continuation bonds on both ends of the structure.
- Put square brackets through the continuation bonds.
- Put 'n' at the bottom right of the square brackets.

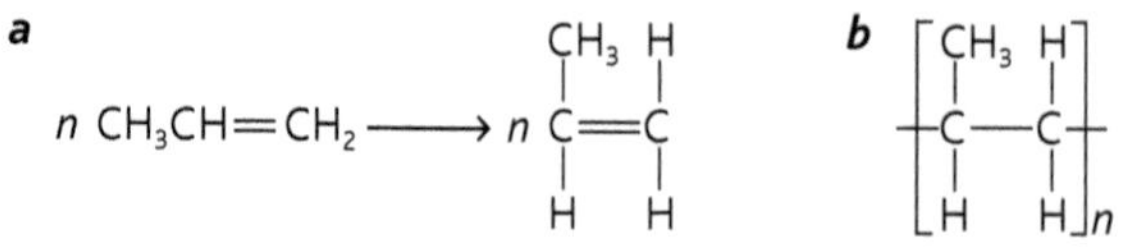

Figure 6.3.1 *Drawing polypropene;* **a** *Rearranging the chain;* **b** *The polymer*

Condensation polymers e.g. a polyester

To draw the structure of the polymer:

- Draw the structure of the monomers and identify the molecule that will be eliminated, e.g. for —COOH and —OH in the molecules H_2O is eliminated.
- Remove an OH from the —COOH and an H from the —OH to make an ester link.

- Put continuation bonds, square brackets and '*n*' around the repeat unit.

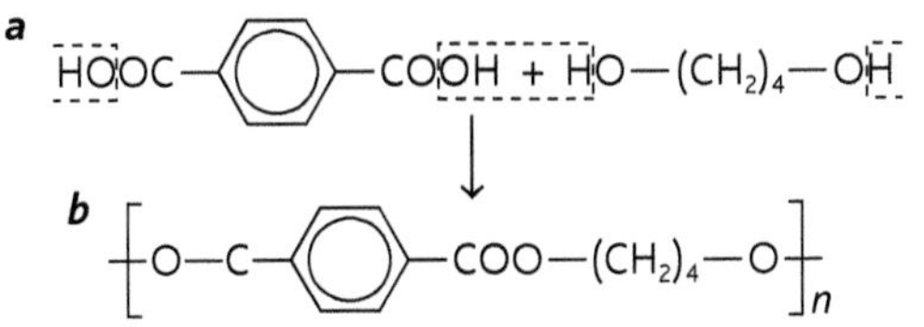

Figure 6.3.2 *Drawing a polyester; **a** The monomers – identifying the atoms eliminated; **b** The polymer*

From polymer to monomer

Addition polymers, e.g. poly(phenylethene)

To draw the structure of the monomer:

- Identify the repeating unit in the polymer and draw this without the brackets.
- Remove the continuation bonds.
- Make the single bond between the carbon atoms in the middle of the repeating unit into double bonds.

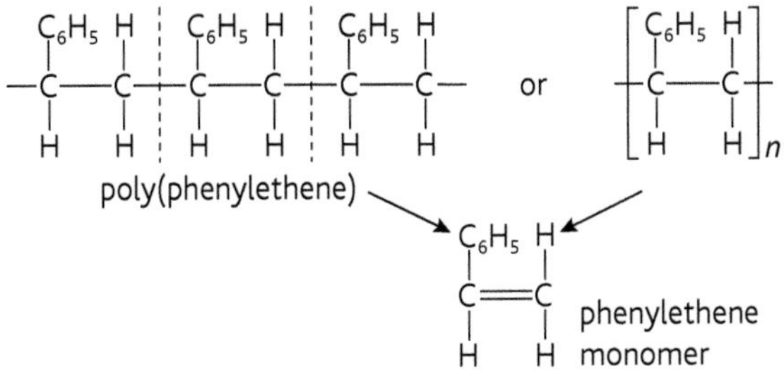

Figure 6.3.3 *Deducing the monomer of poly(phenylethene)*

Condensation polymers

To draw the structure of the monomer:

- Identify the repeating unit in the polymer and 'break' the bonds as follows:

- Add back OH to the C=O group and H to the O or NH group.
- Make the single bond between the carbon atoms in the middle of the repeating unit into double bonds.

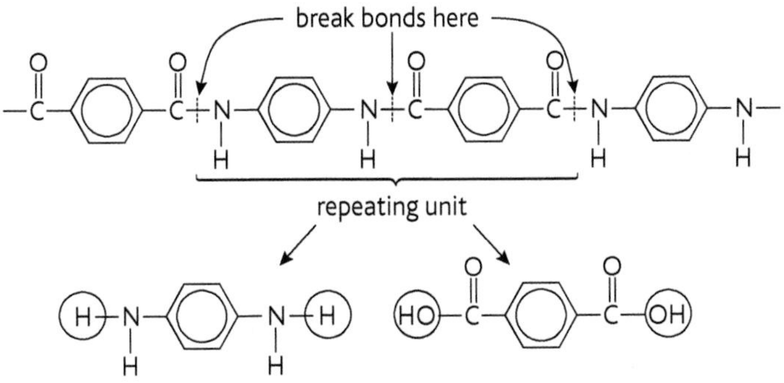

Figure 6.3.4 *Deducing the monomers of a polyamide. The added OH and H groups are shown circled.*

Did you know?

Condensation polymers do not always have to be formed from two different monomers. Nylon-6 can be formed by heating a six-sided ring compound called caprolactam with a little water. In this case the repeating unit of the nylon is $-CONH(CH_2)_5-$.

Key points

- We can draw the simplified formula for a polymer by showing repeating units with square brackets and continuation bonds.

- The structure of a monomer can be deduced from the repeating unit of the polymer by (i) converting single bonds between carbon atoms of the repeating unit in an addition polymer to double bonds, or (ii) by adding OH or H to the end of each repeating unit for a condensation polymer.

6.4 Proteins

Learning outcomes

On completion of this section, you should be able to:

- identify proteins as naturally-occurring macromolecules

- understand that amino acids are the units that condense to form proteins.

Polypeptides

Amino aids can react with each other to form peptides and proteins by condensation reactions. The acidic —COOH group in one amino acid molecule reacts with the basic —NH$_2$ group in another amino acid molecule. The —CO—NH— group formed is called an amide link or a peptide link. When two amino acids react like this a dipeptide is formed and a molecule of water is eliminated. When many amino acids condense a polypeptide is formed.

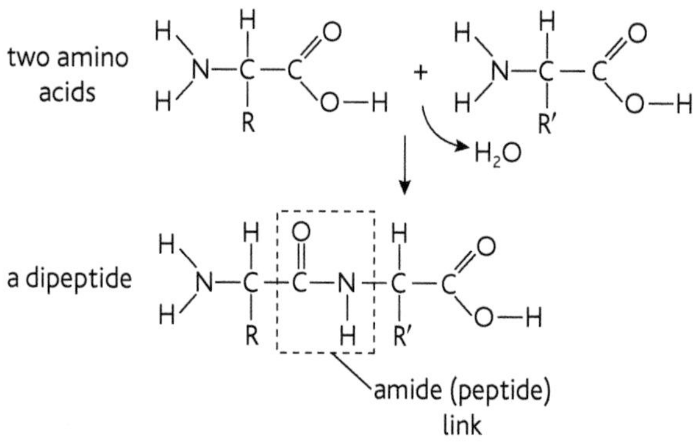

Figure 6.4.1 *The formation of a dipeptide with the elimination of a water molecule*

The C—N bond in the amide link does not rotate freely because of the conjugative (resonance) effect (see Section 5.3), but the C—C bonds either side of this link can rotate. The amide (peptide) link also occurs in polyamides such as nylon, where two different monomers usually condense. In naturally occurring peptides and proteins, however, the 'monomers' can be any of 20 naturally occurring amino acids.

In the laboratory we can make polymers from one type of amino acid such as poly(alanine). Our body, however, makes polypeptides by the stepwise addition of various amino acids, one at a time.

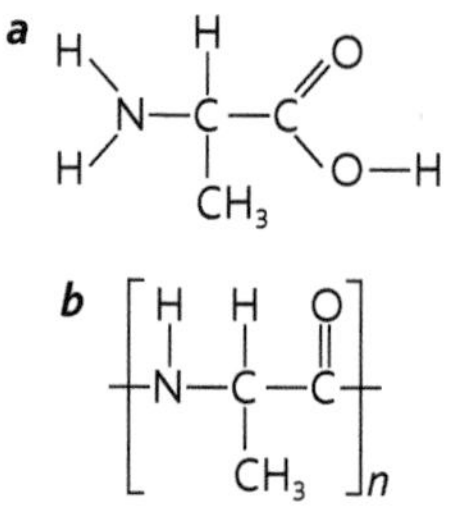

Figure 6.4.2 *The formation of poly(alanine); **a** The alanine monomer; **b** poly(alanine)*

Proteins

Proteins are natural polymers made from 20 naturally-occurring amino acids. Protein is found in muscle, hair, skin, blood and antibodies. Some hormones and all enzymes are proteins. There are thousands of different types of protein. Each of these has a specific sequence of amino acids. Proteins may contain 500 to several thousand amino acids in a particular sequence. The sequence of amino acids along the protein chain called the primary structure of the protein. When amino acids are part of a protein chain we call them amino acid residues.

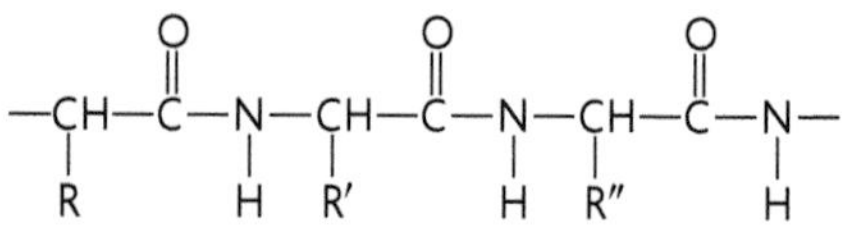

Figure 6.4.3 *Part of the primary structure of a protein. R, R' and R" represent different amino acid side chains*

Did you know?

It can be time-consuming writing out the full chemical names of amino acids, so their common names are often used e.g. glycine rather than 2-aminoethanoic acid. Biochemists use a shorthand way of writing the names of amino acids e.g. Gly is glycine, Ala is alanine, Pro is proline. This shorthand is often the first three letters of the common name of the amino acid.

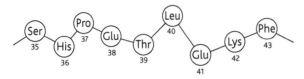

Figure 6.4.4 *Part of the amino acid sequence of the blood pigment myoglobin from a sperm whale. The letters in each circle are shorthand for particular amino acids. The numbers represent the position of the amino acid residues in the chain.*

☑ Exam tips

You should be able to recognise the repeating unit in a protein as —NH—CH(R)—CO— or —CH(R)—CO—NH—. This unit repeats along the chain. R represents one of 20 different side chains.

In living organisms, enzymes catalyse the condensation of amino acids into proteins by a complex series of reactions. A dipeptide is first formed, then a tripeptide (containing three amino acids residues), then a tetrapeptide (containing four amino acid residues) and so on. Some proteins contain more than one chain. The individual chains are called polypeptides.

Did you know?

Wool is a protein fibre with a helical structure joined by a regular arrangement of hydrogen bonds.

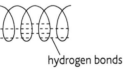

hydrogen bonds

If wool is washed at too high a temperature, the hydrogen bonds break. The clothes may then lose their shape because the hydrogen bonds reform in a less regular way.

Key points

- Proteins are naturally occurring polymers.
- The 'monomers' for proteins are amino acids.
- Proteins are formed by sequential condensation reactions.
- The linkage in proteins is the amide (peptide) link.

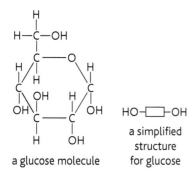

a glucose molecule   a simplified structure for glucose

Figure 6.5.1 *A glucose molecule*

Carbohydrates

Carbohydrate means carbon with water. So the general formula for most simple carbohydrates is $C_x(H_2O)_y$. For example, the molecular formula for glucose is $C_6H_{12}O_6$. Even simple carbohydrates such as glucose are quite complex molecules.

Many carbohydrates contain several —OH groups. For the purposes of understanding the polymerisation of carbohydrates we can write a simple carbohydrate in a simplified form as HO—■—OH.

- Simple sugars such as glucose and fructose are called monosaccharides because they contain one sugar unit (mono means one and saccharide means sugar).
- Sugars containing two simple sugar units, e.g. sucrose are called disaccharides.
- Sugars containing many simple sugar units are called polysaccharides.

Formation of polysaccharides

The polymerisation of monosaccharides such as glucose to form polymers (**polysaccharides**) is an example of condensation polymerisation. Water is eliminated. A simplified diagram of this process is shown in Figure 6.5.2.

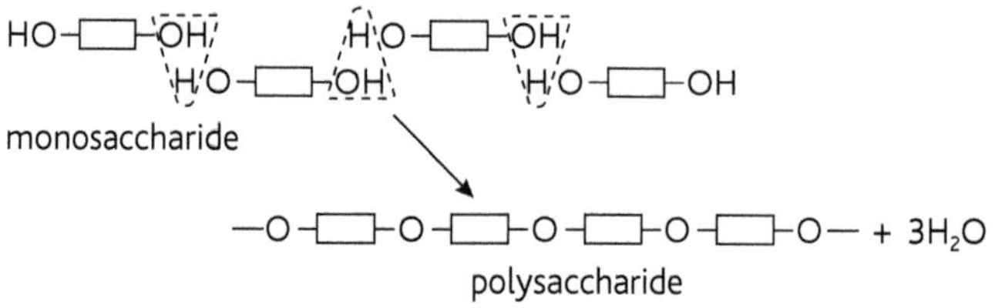

Figure 6.5.2 *Monosaccharide molecules condensing to form a polysaccharide. The monosaccharide is shown as HO—■—OH.*

The C—O—C linkage in these sugar polymers is called a glycosidic link. The empirical formula for a polysaccharide made from glucose is $(C_6H_{10}O_5)_n$ i.e. glucose with water removed.

Naturally-occurring polysaccharides

Polysaccharides are found:

- as storage carbohydrate (starch in plants and glycogen in animals)
- in plant cell walls (cellulose)
- as a 'glue' between plant cell walls (pectin).

These polysaccharides are made in living organisms using enzymes as catalysts.

They are all made by condensation polymerisation.

Starch

Starch provides us with most of the carbohydrate in our diet. It is a polymer of hundreds of glucose units. It can form chains or branched chains. The glucose monomers polymerise by the —OH groups at the

1- and 4-positions condensing and eliminating water. Note that the group in position 6 is always on the same side of the chain. Polymerisation can also happen between the —OH groups at the 1- and 6-positions. This leads to a branched chain form of starch.

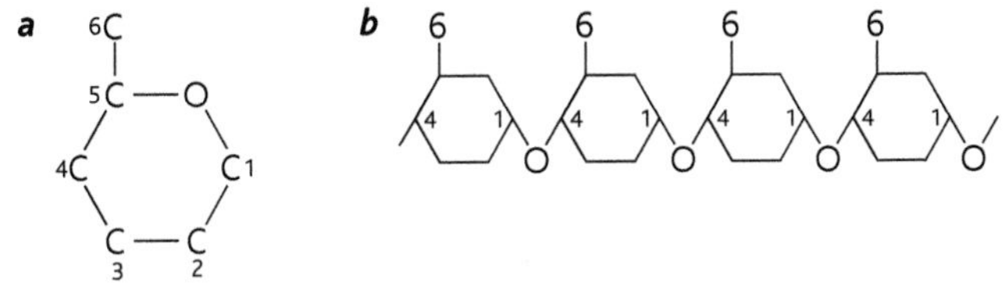

Figure 6.5.3 *The simplified structure of starch;* **a** *The numbering of the carbon atoms in a glucose molecule;* **b** *A simplified diagram of part of a starch chain showing the α-1,4-linkages*

Did you know?

Glucose has several chiral centres (see Section 1.6). There are several optical isomers of glucose. The form used for starch synthesis is called α-D-glucopyranose. In addition glucose exists in a chain form as well as a ring form.

Cellulose

Cellulose is also made from glucose monomers. The enzymes that are responsible for this polymerisation act on a different isomer of glucose called β-glucose. In this isomer, the —H and —OH atoms in position 4 of the ring have a different position in space. In cellulose, the group in position 6 is not always on the 'same side' of the chain (see Figure 6.5.4).

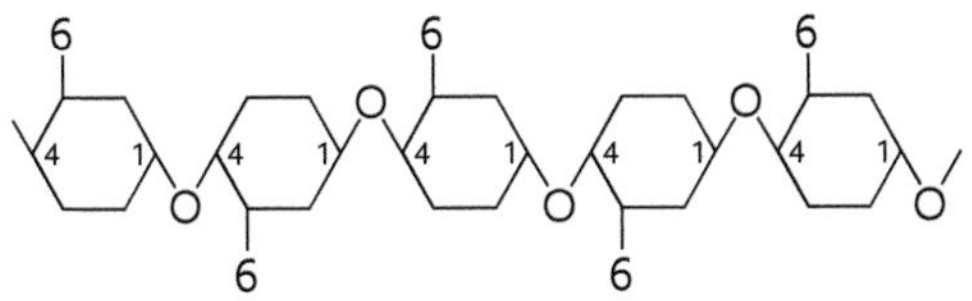

Figure 6.5.4 *The simplified structure of cellulose. Note that the β-1,4-linkages cause the group in the 6 position to alternate in the chain.*

Pectin

Pectin is found between the walls of plant cells. It is used commercially to make jam set. The monomer for the formation of pectin is methylglucuronic acid. This is similar to glucose except that the —CH_2OH group at position 6 in glucose is replaced by a —$COOCH_3$ group. The link is a β-1,4-link as in cellulose.

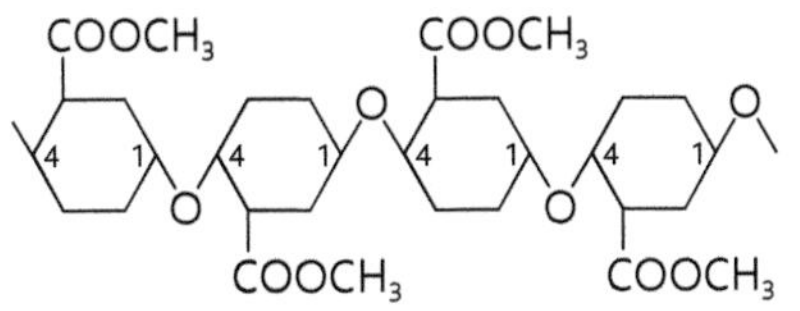

Figure 6.5.5 *The simplified structure of pectin. Note that the β-1,4-linkages cause the group in the 6-position to alternate in the chain.*

Did you know?

Scientists have found a way of producing hydrogen from cellulose using microorganisms. If this method can be modified on an industrial scale, the hydrogen produced could be used as a fuel for cars.

Key points

- Polysaccharides are naturally-occurring polymers made from monomers of simple sugars.
- The monomers in polysaccharides are usually glucose or glucose esters.
- The linkage in polysaccharides between the sugar units, —O— is called a glycosidic link.
- Starch and cellulose are glucose polymers.
- Pectin is a polymer of methylglucuronic acid.

Exam-style questions – Module 1

Answers to all exam-style questions can be found on the accompanying CD.

Multiple-choice questions

1 Which types of species best describe the products of a heterolytic fission of a covalent bond in a diatomic molecule?

   A electrophiles and nucleophiles
   B atoms and free radicals
   C electrophiles and free radicals
   D nucleophiles and atoms

2 What is the IUPAC name for the compound below?

   $CH_3C(CH_3)ClCH_2CH_2CH_3$

   A 2,2-chloromethyl pentane
   B 2-chloro-2-methyl pentane
   C 2- methyl-2-chloro pentane
   D 4-chloro-4-methyl pentane

3 Which of the following are properties of a homologous series?

   i   Members can be represented by a general formula.
   ii   Members possess similar chemical properties.
   iii Members possess the same empirical formulae.
   iv   Members differ by a CH_2 group.

   A i, ii, iii
   B i, ii, iv
   C i, ii
   D ii, iv

4 Which of the following reagents can be used to distinguish between compounds **A** and **B**?

   $CH_3\,CH_2\,C\!\!\begin{array}{c}O\\H\end{array}$ $CH_3\,C\!\!\begin{array}{c}O\\\end{array}\!CH_3$

   A   **B**

   i   acidified potassium manganate(VII)
   ii   Tollen's reagent
   iii Brady's reagent
   iv   NaCN (aq) and dilute HCl

   A i and ii
   B i and iii
   C ii and iv
   D iii and iv

5 On complete combustion a hydrocarbon produced 0.264 g of carbon dioxide and 0.054 g of water.

   Which of the following compounds correctly satisfies this analysis?

   A $CH_2\!-\!CH\!=\!CH_2$

   B

   C $CH_2\!=\!CH_2$

   D CH_3

Structured questions

6 a Explain the difference between i empirical and ii structural formulae.   [2]

   b A hydrocarbon, **P**, present as a minor constituent of crude oil has the formula C_4H_8.
      i   Write the displayed formulae of **three** isomers of **P**.   [3]
      ii One of these isomers, when treated with hot acidified potassium manganate(VII), produced two compounds, **Q** (C_3H_6O) and **R** (CH_2O). Show the steps involved in the mechanism of the reaction between **Q** and HCN. Use curved arrows to show the movement of electrons.   [5]

   c The product formed by the reaction in **b ii** exhibits isomerism.
      i   State the type of isomerism and its characteristic feature.   [2]
      ii   Draw the displayed formulae of the isomers involved.   [2]
      iii State the property which allows them to be identified.   [1]

7 a i   Explain the meaning of addition polymerisation.   [2]
      ii Phenyl ethene, **A**, forms a polymer polyphenylethene (polystyrene).

          $CH\!=\!CH_2$

          A

          Draw the structure of part of the polymer involving three repeating units.   [2]

   b Define the term 'condensation polymerisation'. [2]

   c The structure, **B**, represents the repeating unit of a polymeric substance.

      B

      i   Name the link present in **B**.   [1]
      ii Deduce the structural formulae of the monomers used to form the polymer.   [4]

c Using your knowledge of their reaction to heat, explain the difference between i thermoplastic and ii thermosetting polymers. [2]

d List two classes of naturally occurring macromolecules. [2]

8 Combustion of 1.00 g of an organic compound, **R**, gave 2.20 g of carbon dioxide and 1.21 g of water. **R** contains carbon, hydrogen and oxygen only. 1.00 g of **R** occupied a volume of 373 cm³ at s.t.p.

a Calculate the empirical formula of **R**. [5]

b Deduce the molecular formula of **R**. [2]

c **R** reacts with the following reagents:

- methanoic acid
- conc. sulphuric acid
- alkaline iodine solution.

i State the name of **R** and give reasons for your conclusion. [2]

ii Write equations to represent the reactions of **R** with the first two reagents listed above. [4]

iii State what would be observed and draw the displayed formula of the organic product formed with the final reagent listed above. [2]

9 An ester of a monohydric alcohol and a monocarboxylic acid is completely hydrolysed by aqueous sodium hydroxide. 1.63 g of the ester required 2.2×10^{-2} moles of sodium hydroxide.

a i Calculate the molecular mass of the ester. [2]

ii Suggest the name and draw the structural formula of the ester. [2]

iii Draw the displayed formula of one other ester with the same molecular formula. [1]

b i State the type of reaction involved if the ester in **a** above was replaced by an oil or fat. [1]

ii State the industrial significance of this type of reaction. [1]

c The ester **A**, below, reacts with ethanol according to the following equation:

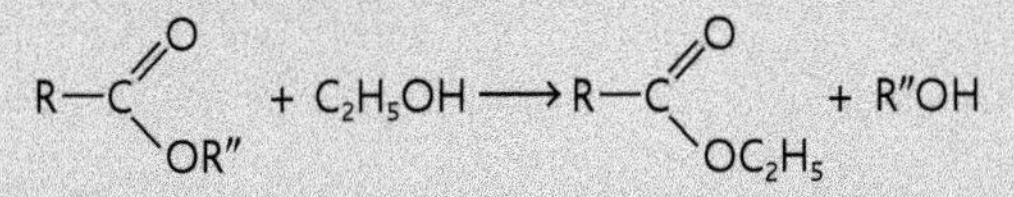

$$R-C\underset{OR''}{\overset{O}{\diagdown}} + C_2H_5OH \longrightarrow R-C\underset{OC_2H_5}{\overset{O}{\diagdown}} + R''OH$$

A

i Give the name of the process involved in this reaction. [1]

ii Explain its industrial importance. [2]

d When a mixture of ethane and chlorine is irradiated with ultraviolet light a number of organic products are formed.

i Give the name of the process involved in this reaction. [1]

ii One of the products is chloroethane. Explain the mechanism involved in its production. Use fish hook notation to describe the movement of electrons. [4]

10 Eugenol is an aromatic liquid which can be extracted from cloves. Its structural formula is represented below:

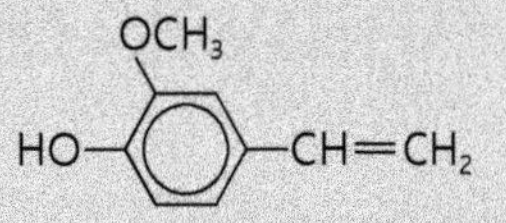

a Describe how you would expect eugenol to react with the following reagents, drawing structural formulae of the products and giving any relevant observations.

i sulphur dichloride oxide, $SOCl_2$ [2]

ii $Br_2(l)/CCl_4$ [2]

iii $Br_2(aq)$ [3]

b Describe a simple laboratory test to distinguish between the following pairs of compounds:

i 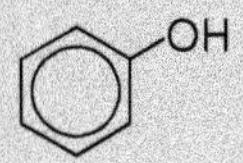 and CH_3CO_2H [3]

ii 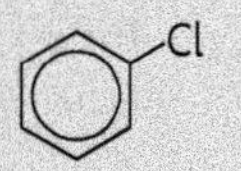 and CH_3CH_2Cl [3]

iii 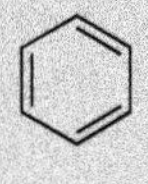 and 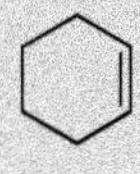 [3]

7.1   Analysis of scientific data

Accuracy and precision

When we carry out chemical experiments such as titrations, gravimetric analysis or collecting data relevant to reaction rates, we need to know whether our measurements are accurate or not. **Accurate measurements** are very close to the true value.

You can get accurate data by:

- repeating the measurements many times
- repeating the measurements using different instruments
- using measuring instruments which are very accurate
- using measuring instruments carefully.

Precision means how close the measurements are to each other. If the measurements are very close to each other, they are **precise**.

An idea of the difference between accuracy and precision is shown in Figure 7.1.1, where the results of different titres are shown.

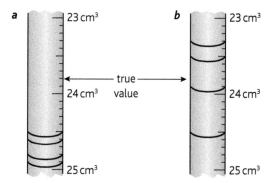

Figure 7.1.1 *The black lines across the burette in **a** and **b** show four different burette readings for the same experiment; **a** The results are precise but not accurate; **b** The results are accurate (because the average is close to the true value) but not precise.*

☑ Exam tips

When thinking about the difference between accuracy and precision, the idea of shooting at a target may help.

In **a** the shots are precise but not accurate. In **b** the shots are accurate but not precise.

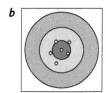

A set of repeat readings in chemistry should have a mean (average value) close to the true value and be precise.

Mean value

The **mean** is the average of the numbers taken from the data in a number of identical experiments. For example:

- A fuel is used to heat a fixed volume of water.
- The temperature rise is measured after a set time.
- The experiment is repeated five times.

The results for the experiments are:

Experiment	1	2	3	4	5
Temperature rise/ °C	10.1	12.0	14.2	13.5	12.7

The mean is $\dfrac{10.1 + 12.0 + 14.2 + 13.5 + 12.7}{5} = \dfrac{62.5}{5} = 12.5\,°C$

Standard deviation

Standard deviation is a measure of how spread out the numbers are from the mean. A low value shows that the data points are close to the mean. A high value shows that the data points are spread out over a wider range. The standard deviation is only 'significant' if it falls outside the normal variation expected from the measuring instruments used. In chemistry, we use the 'sample standard deviation'. The equation we use is:

$$S_N = \sqrt{\frac{\Sigma(x-\bar{x})^2}{n-1}}$$

S_N is the sample standard deviation

x is each individual piece of data

$\bar{x}$ is the mean

Σ is the sum of $(x - \bar{x})^2$

n is the number of individual pieces of data.

The standard deviation has the same units as the data used.

Worked example

In a titration experiment to find the concentration of an alkali, the values of four titres are: $19.6\,cm^3$; $20.0\,cm^3$; $20.2\,cm^3$; $19.4\,cm^3$. Calculate the standard deviation.

Step 1 Find the mean: $\dfrac{19.6 + 20.0 + 20.2 + 19.4}{4} = 19.8\,cm^3$

Step 2 Find the sum of the squares of the differences from the mean:

$(19.6 - 19.8)^2 + (20.0 - 19.8)^2 + (20.2 - 19.8)^2 + (19.4 - 19.8)^2$

$0.04 + 0.04 + 0.16 + 0.16 = 0.40$

Step 3 Divide by $n - 1$, i.e. $0.40 \div 3 = 0.13$

Step 4 Take the square root: $\sqrt{0.13} = 0.36\,cm^3$

This could be considered quite a high standard deviation as we can read the burette to an accuracy of at least $0.1\,cm^3$ and perhaps to $0.05\,cm^3$. A second example is given in Section 8.1.

Key points

- The mean is the average of a sample of data.

- Standard deviation is a measure of how far the data deviates from the mean.

- Precision refers to how closely the data values are grouped together – the closer the values, the greater the precision.

- Accuracy refers to the closeness of the data values to the true value.

Learning outcomes

On completion of this section, you should be able to:

- assess the degree of uncertainty in measurements associated with pieces of laboratory apparatus
- select appropriate apparatus to make measurements depending on the degree of accuracy required.

✓ Exam tips

1 When using an accurate balance, inaccuracies in weighing can be caused by air draughts or greasy finger marks on the weighing bottle.

2 It is bad practice to try to weigh out an exact amount, e.g. 1.30 g of solid to make an exact solution, e.g. 0.1 mol dm⁻³.

Errors in practical chemistry

There are three main causes of errors in practical chemistry:

- mistakes in calculations, including mistakes with significant figures
- faults in laboratory equipment
- limitations of the apparatus used.

Weighing

For making up small quantities e.g. $500\,cm^3$ of solutions for titrations, we need to weigh to an accuracy of $\pm 0.01\,g$. For accurate gravimetric work an accuracy of $\pm 0.001\,g$ or $\pm 0.0001\,g$ is needed. It is always the loss of mass that is measured i.e.

(mass of weighing bottle + chemical) – (mass of weighing bottle alone)

Did you know?

The earliest record of a balance dates from over 4000 years ago from the Indus Valley in present day Pakistan. Simple beam balances for accurate weighing have been present in chemistry laboratories since the 19th century. The modern day balance for accurate work should really be called an analytical scale rather than an analytical balance. This is because it measures force rather than gravitational mass.

Volumes and temperatures

Pieces of laboratory glassware have calibration marks which guarantee maximum errors. A calibration mark is a line on the glassware that shows a particular value of volume, e.g. $100\,cm^3$. These volumes are usually measured at a particular temperature (usually $20\,°C$). The table shows some typical errors for some pieces of class B titration apparatus used in most schools.

Apparatus	Maximum error
1 dm³ standard flask	±0.8 cm³
250 cm³ standard flask	±0.3 cm³
50 cm³ burette	±0.1 cm³ between any two marks
25 cm³ volumetric pipette	±0.06 cm³

Pieces of glassware such as large measuring cylinders are much more inaccurate, e.g. $\pm 1\,cm^3$. They should not be used for measuring volumes accurately. The graduation marks on beakers are even more inaccurate.

If temperatures are to be measured, the accuracy of the thermometer may play a part in the overall accuracy of the experiment. Some thermometers are available which read to $\pm 0.01\,°C$, but most laboratory thermometers only read to the nearest degree Celsius.

Overall experimental accuracy

Experiments should be designed to get the best accuracy out of the apparatus available.

- Burette error is minimised by having titres that are above $30\,cm^3$, masses measured to the nearest $0.01\,g$ and volumes measured with a volumetric pipette.

- The best accuracy overall in a school laboratory is only likely to be in the order of 0.5–1%. So there is little point in quoting final results with a precision greater than this. For example, quoting a concentration to three significant figures would be incorrect.

- The overall accuracy of the experiment will depend on the piece of apparatus that is least accurate. It is of little value weighing approximately 1 gram of a solid to three decimal places, if the accuracy of the $100\,cm^3$ container you are making a solution of the solid in is accurate to only $1\,cm^3$.

- When making solutions, it is more accurate to make up a large volume of a solution in a volumetric flask than a small volume. Similarly, weighing larger quantities of a substance is more accurate that weighing smaller quantities.

Making up solutions

To prepare a solution of known concentration, we need to weigh out the solute to the required degree of accuracy and use a volumetric flask to prepare the solution.

To make $200\,cm^3$ of a solution of known concentration, the procedure is:

- Tip the solid from the weighing bottle into a $200\,cm^3$ beaker.
- Add about $50\,cm^3$ of pure water.
- Shake well to dissolve the solid.
- Wash out the volumetric flask with a little pure water.
- Pour the solution from the beaker into the volumetric flask using a funnel.
- Wash out the beaker several times with pure water and add the washings to the flask.
- Wash out any liquid remaining in the funnel into the volumetric flask with a little pure water.
- Fill the volumetric flask with pure water to just below the meniscus.
- Add water dropwise until the bottom of the meniscus is on the calibration mark.
- Put the stopper (bung) on the flask and shake gently.

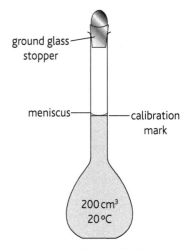

Figure 7.2.1 *A volumetric flask used to make a standard solution*

Key points

- Pipettes, burettes and volumetric flasks have calibration marks with a known maximum error.

- The overall accuracy is dependent on the piece of apparatus that is least accurate.

- In volumetric analysis, weighing and measuring volumes should be made to the appropriate number of significant figures for the overall accuracy required.

7.3 Standards

Learning outcomes

On completion of this section, you should be able to:

- understand the criteria used in selecting primary standards
- identify the use of $NaHCO_3$, Na_2CO_3, KIO_3, $(COOH)_2$ and its salts as primary standards
- understand criteria for the preparation of standard solutions
- understand the use of calibration curves.

Introduction

We compare chemical quantities and measurements in terms of standard values. We have already met terms such as *standard electrode potential* and *standard enthalpy change*. Relative atomic mass is measured on the carbon-12 scale. In addition:

- Standard temperature is $298\,K$.
- Standard pressure is $101\,325\,Pa$.

Primary standards used in titrations

In order to find the concentration of a solution by titration, we need to prepare standard solutions of known concentrations. For example, when titrating an unknown alkali with hydrochloric acid, we need to know the concentration of the acid to at least two significant figures, e.g. $0.014\,mol\,dm^{-3}$. We make sure that the acid has the correct concentration by titrating it with a primary standard.

A **primary standard** for use in titrations is a chemical with the following properties:

- The solid must be able to be obtained to a very high purity.
- It must be stable in air.
- It must be readily soluble in water and form a stable solution.
- It must give reproducible results in a titration.
- It should preferably have a high relative molecular mass.

Primary standards can be used to find the exact concentrations of acid, alkalis, reducing agents or oxidising agents.

Some examples of primary standards

The table shows some primary standards commonly used in the laboratory. All of these standards are available to a high level of purity.

Primary standard	Used to standardise
sodium carbonate, Na_2CO_3	acids
sodium hydrogencarbonate	acids
potassium iodide, KIO_3	sodium thiosulphate
ethanedioic acid (oxalic acid), $(COOH)_2$	bases and some oxidising agents
potassium dichromate(VI), $K_2Cr_2O_7$	reducing agents
sodium chloride, $NaCl$	silver nitrate

Did you know?

For the highest accuracy work, silver of 99.9999% purity is used as a standard. All other standards are calibrated against this.

Standardising solutions used in colorimetry

Colorimetry is an easy and quick way of finding the concentration of coloured solutions (see *Unit 1 Study Guide*, Section 7.1). Figure 7.3.1 shows a simplified diagram of a colorimeter.

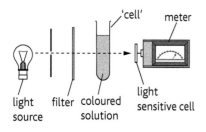

light source   filter   coloured solution   'cell'   meter   light sensitive cell

Figure 7.3.1 *A colorimeter*

The electric current registered on the meter is proportional to the intensity of light falling on the light-sensitive cell. Before using the instrument, it must be **calibrated**. In order to calibrate the colorimeter, we have to see how the meter readings change when different concentrations of solutions are placed in the cell. This is done by:

■ making a set of solutions of known different concentrations by accurate dilution of a standard solution

■ taking the meter readings of each solution.

The procedure is:

■ Put a cell containing the pure solvent used to make the solutions into the colorimeter.

■ Adjust the meter reading to zero.

■ Put a cell containing a solution of known concentration into the colorimeter.

■ Record the meter reading.

■ Repeat these steps for other solutions.

The meter readings are then plotted against the concentrations of the solutions.

For very dilute coloured solutions, the meter reading is likely to be proportional to the concentration of the solution. If the solutions are concentrated, the meter reading may not be proportional to the concentration. We can, however, find the values of the concentration of the coloured solution from the **calibration curve** as shown in Figure 7.3.2.

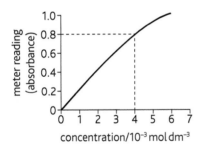

Figure 7.3.2 Calibration curve for a coloured solution. The concentration of the coloured solution at meter reading 0.8 is 4×10^{-3} mol dm^3.

For more information about colorimetry see Section 9.2.

Key points

■ Primary standards are used to calculate the concentration of acids, alkalis or other substances used in titrations.

■ Na_2CO_3 is used as a primary standard to standardise acids.

■ KIO_3 can be used to standardise sodium thiosulphate.

■ $(COOH)_2$ can be used to standardise bases.

■ $K_2Cr_2O_7$ can be used to standardise reducing agents.

■ When making up solutions of known concentration, flasks and balances measuring to the required degree of accuracy should be selected.

■ When measuring concentrations using a colorimeter, reference should be made to a calibration curve for the instrument.

8.1   Principles of titrations

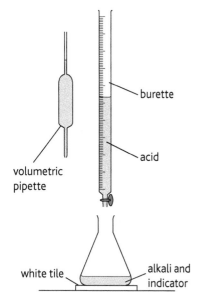

Figure 8.1.1 *The apparatus used in an acid–alkali titration*

Carrying out a titration

A titration is used to determine the amount of substance present in a solution of unknown concentration. This is the procedure for determining the concentration of a solution of alkali:

▨ Fill a burette with acid of known concentration (after washing it with the acid).

▨ Record the initial burette reading.

▨ Put a known volume of alkali into the flask using a volumetric pipette.

▨ Add an acid–base indicator to the alkali in the flask.

▨ Add the acid slowly from the burette until the indicator changes colour (end point).

▨ Record the final burette reading (final – initial burette reading is called the titre).

▨ Repeat the process until two or three successive titres differ by no more than $0.10\,cm^3$.

Titres and standard deviation

For our results, we take successive titres which differ by no more than $0.10\,cm^3$. This gives us a standard deviation which is very low so we can be sure that the experiment is very accurate. For example in the table below, the 4th, 5th and 6th titres would be selected. The mean of these figures is 32.30.

Rough titre	2nd titre	3rd titre	4th titre	5th titre	6th titre
$32.95\,cm^3$	$32.85\,cm^3$	$32.00\,cm^3$	$32.25\,cm^3$	$32.35\,cm^3$	$32.30\,cm^3$

The standard deviation (see page 71) for these three titres is:

$$\sqrt{\frac{(32.25-32.30)^2 + (32.35-32.30)^2 + (32.30-32.30)^2}{2}}$$

The standard deviation for the 4th to 6th titres is $0.05\,cm^3$, which is very low. If we took the 2nd to 6th titres, however, the standard deviation is much higher: $0.31\,cm^3$.

Titrimetric technique

Using a pipette

A volumetric pipette is designed to deliver a fixed volume of liquid when it is filled to its calibration mark. When using a pipette:

▨ Have the solution that is to be used in a beaker.

▨ Using a pipette filler, wash out the pipette with the solution by sucking some of it up, then letting it drain out into the sink.

▨ Fill the pipette again so the liquid level is just above the calibration mark.

- Remove the pipette from the beaker.
- Bring the solution level down to the calibration mark so that the meniscus just touches this mark.
- Run the contents of the pipette into the clean titration flask (or a flask washed with pure water).
- Allow the pipette to drain completely by keeping the tip in contact with the side of the flask after the solution has been delivered.

Using a burette

When using a burette:

- Rinse the burette with the solution to be used in it then allow the solution to drain through the tip of the burette.
- Clamp the burette vertically and put a beaker beneath the burette.
- Using a funnel, add a little of the solution to be used to the burette with the tap open. Close the tap while there is still liquid in the burette and make sure that there are no air bubbles in the tip of the burette.
- Fill the burette. Remove the funnel.
- Adjust the level of the meniscus to a definite graduation (calibration) mark. Make sure that you take the reading from the bottom of the meniscus (see Figure 8.1.2).
- Place the titration flask and its contents below the burette.
- Turn the burette tap with your left hand (or right hand if you are left-handed). This leaves the other hand free to shake the flask.
- Run in the solution from the burette while shaking the flask from side to side.
- When doing accurate (rather than rough) titrations, you must add the solution from the burette one drop at a time when the end point is approached. This prevents you overshooting the endpoint and getting too high a value for the titre.

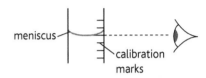

Figure 8.1.2 *Reading a burette. Your eye should be level with the bottom of the meniscus.*

Key points

- Acid–base titrations are carried out using an indicator which changes colour rapidly at the end point.
- When processing titration results, the values selected should be from two or three successive titres whose values are no more than $0.10\,cm^3$ apart.

On completion of this section, you should be able to:

- understand the basic principles of back titrations

- perform calculations based on back titrations.

Back titrations

It is sometimes easier to do a titration in reverse. This is called a **back titration**. In a back titration, a known amount of a standard reagent is added in excess to the solution whose concentration we wish to find. The excess reagent is then titrated with a standard solution. Back titrations are useful when:

- the reaction is very slow

- the substance to be titrated is an insoluble solid, e.g. calcium carbonate

- one of the reactants is volatile, e.g. ammonia

- the end point of the titration is difficult to observe.

Calculating an unknown from a back titration

General procedure

1 The substance of unknown concentration is reacted with an excess of another reagent whose amount (in moles) is known.

2 A titration is carried out to find the amount of added reagent which is in excess.

3 The number of moles of excess reagent calculated from the titration is subtracted from the number of moles of reagent added originally.

Determining the mass of calcium carbonate present in a sample of marble

- Put a small sample of marble in a titration flask.

- Add $50\,cm^3$ of $0.25\,mol\,dm^{-3}$ hydrochloric acid to the marble (an excess). Use a volumetric pipette for this.

- Shake the contents of the flask until all the calcium carbonate has reacted.

$$2HCl(aq) + CaCO_3(s) \rightarrow CaCl_2(aq) + CO_2(g) + H_2O(l)$$

- Add more hydrochloric acid of known concentration and volume if the reaction is not complete.

- Titrate the excess hydrochloric acid with a standard sodium hydroxide solution using a suitable acid–base indicator.

Worked example 1

A sample of $0.300\,g$ of limestone reacts completely with $50.0\,cm^3$ of $0.250\,mol\,dm^{-3}$ hydrochloric acid (an excess). It required $35.5\,cm^3$ of $0.200\,mol\,dm^{-3}$ sodium hydroxide to neutralise the excess hydrochloric acid. Calculate the mass of calcium carbonate in the sample of limestone assuming that this is the only carbonate present.

Step 1: Calculate the moles of NaOH

$$\text{mol NaOH} = \frac{35.5}{1000} \times 0.200 = 7.10 \times 10^{-3}\,mol$$

Step 2: Calculate the moles of HCl which react with this.

$$NaOH(aq) + HCl(aq) \rightarrow NaCl(aq) + H_2O(l)$$

mol HCl = $7.10 \times 10^{-3}\,mol$ (since 1 mol of HCl reacts with 1 mol NaOH)

✓ Exam tips

Make sure that you revise the use of:

1 $\text{moles} = \dfrac{\text{mass (g)}}{\text{molar mass (g mol}^{-1})}$

2 $\text{concentration (mol dm}^{-3})$
$= \dfrac{\text{moles}}{\text{volume (dm}^3)}$

Take account of the stoichiometry of the equation.

Step 3: Calculate the number of moles of HCl initially added.

$$\text{Moles HCl} = \frac{50.0}{1000} \times 0.250 = 0.0125\,\text{mol}$$

Step 4: Calculate the number of moles of HCl that reacted with the $CaCO_3$

$$0.0125 - 7.10 \times 10^{-3} = 5.40 \times 10^{-3}\,\text{mol}$$

Step 5: Calculate the number of moles of calcium carbonate which react

$$2HCl(aq) + CaCO_3(s) \rightarrow CaCl_2(aq) + CO_2(g) + H_2O(l)$$

$$= \frac{5.40 \times 10^{-3}}{2} = 2.70 \times 10^{-3}\,\text{mol}$$

(since 2 mol of HCl reacts with 1 mol $CaCO_3$)

Step 6: Calculate the mass of $CaCO_3$ (molar mass = $100\,\text{g mol}^{-1}$).

$$2.70 \times 10^{-3} \times 100 = 0.270\,\text{g}$$

Worked example 2

A solution containing $45.0\,\text{cm}^3$ of $0.200\,\text{mol dm}^{-3}$ hydrochloric acid was added to a $20.0\,\text{cm}^3$ sample of aqueous ammonia of unknown concentration. The hydrochloric acid was in excess. The excess hydrochloric acid was titrated with $0.0500\,\text{mol dm}^{-3}$ aqueous sodium carbonate. It required $27.5\,\text{cm}^3$ of the sodium carbonate solution to neutralise the hydrochloric acid. Calculate the concentration of the aqueous ammonia.

Step 1: Calculate the moles of sodium carbonate.

$$\text{Moles Na}_2CO_3 = \frac{27.5}{1000} \times 0.050 = 1.375 \times 10^{-3}\,\text{mol}$$

Step 2: Calculate the moles of HCl which react with this.

$$Na_2CO_3(aq) + 2HCl(aq) \rightarrow 2NaCl(aq) + H_2O(l) + CO_2(g)$$

mol HCl = 2.75×10^{-3} mol (since 2 mol of HCl reacts with 1 mol Na_2CO_3)

Step 3: Calculate the number of moles of HCl initially added.

$$\text{Moles HCl} = \frac{45.0}{1000} \times 0.200 = 9.00 \times 10^{-3}\,\text{mol}$$

Step 4: Calculate the number of moles of HCl that reacted with the NH_3

$$9.00 \times 10^{-3} - 2.75 \times 10^{-3} = 6.25 \times 10^{-3}\,\text{mol}$$

Step 5: Calculate the number of moles of ammonia which react.

$$HCl(aq) + NH_3(g) \rightarrow NH_4^+(aq) + Cl^-(aq)$$

$= 6.25 \times 10^{-3}$ (since 1 mol of HCl reacts with 1 mol NH_3)

Step 6: Calculate the concentration of the aqueous ammonia.

$$6.25 \times 10^{-3} \times \frac{1000}{20.0} = 0.313\,\text{mol dm}^{-3}\ (\text{to 3 s.f.})$$

Key points

- In a back titration, excess of one of the reagents of known concentration and volume is added to the reagent under test. The excess reagent is then titrated.

- The amount of substance consumed in a back titration = (moles substance originally added – moles substance calculated from the titration).

- In titrations, calculations involve use of the relationship:

  concentration (in mol dm^3) = amount (in mol) ÷ volume (in dm^3)

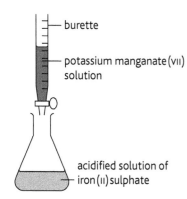

Figure 8.3.1 *The titration of aqueous iron(II) sulphate with potassium manganate(VII)*

Introduction

Redox titrations are used to calculate the concentrations of oxidising or reducing reagents. The titration is carried out in a similar manner to acid–base titrations. The indicator used can be:

- one of the reactants which acts as an indicator because it exhibits a particular colour when the reaction is complete and it is in excess
- an added redox indicator which changes colour when the reaction is complete.

Potassium manganate(VII) as a redox indicator

Potassium manganate(VII) is a good oxidising agent. It can therefore be used to calculate the concentration of reducing agents such as Fe^{2+} ions, H_2O_2 (hydrogen peroxide) or ethanedioic acid (oxalic acid).

In acidic solution, iron(II) ions react with the manganate(VII) ions, MnO_4^-, in potassium manganate(VII) according to the equation:

$$5Fe^{2+}(aq) + MnO_4^-(aq) + 8H^+(aq) \rightarrow 5Fe^{3+}(aq) + Mn^{2+}(aq) + 4H_2O(l)$$

pale green   deep purple   yellow   very pale pink

Figure 8.3.1 shows the apparatus used for this redox titration.

In this titration:

- Potassium manganate(VII), $KMnO_4$, is added gradually from the burette to the acidified solution of iron(II) sulphate in the flask.
- When the potassium manganate(VII) is added to the flask it loses it purple colour. This is because the purple MnO_4^- ions are changed to almost colourless Mn^{2+} ions by reaction with the Fe^{2+} ions.
- When just enough potassium manganate(VII) has been added to the flask to react with all the Fe^{2+} ions, the addition of a further drop of potassium manganate(VII) results in the solution in the flask turning purplish-pink. This is the end point of the titration. Note that an indicator is not added, because the potassium manganate(VII) is self indicating in this redox reaction.

Permanganate titration: worked example

Iron tablets contain Fe^{2+} ions as iron(II) sulphate. One iron tablet is dissolved in excess sulphuric acid and made up to $100\,cm^3$ in a volumetric flask. A sample of $10.0\,cm^3$ of this solution was titrated with $0.001\,00\,mol\,dm^{-3}$ potassium manganate(VII), $KMnO_4$. It required $22.5\,cm^3$ of potassium manganate(VII) to react completely with the Fe^{2+} ions. Calculate the mass of iron(II) sulphate ($M = 151.9$) in one iron tablet.

Step 1: Calculate moles of the $KMnO_4 = \dfrac{22.5}{1000} \times 0.001\,00$

$$= 2.25 \times 10^{-5}\,mol$$

Step 2: Calculate moles of Fe^{2+} using the stoichiometric equation (see above).

$$2.25 \times 10^{-5} \times 5\,mol = 1.125 \times 10^{-4}\,mol\ Fe^{2+}$$

Step 3: Calculate the mass of iron(II) sulphate in the flask (from 1 tablet)

$1.125 \times 10^{-4} \times 100/ 10 = 1.125 \times 10^{-3} \, mol$

We divided by 10 because $10 \, cm^3$ of the $100 \, cm^3$ were taken for titration.

Mass of iron(II) sulphate $= 1.125 \times 10^{-3} \times 151.9 = 0.171 \, g$

Potassium dichromate(VI) titrations

The method is similar to the method used for a potassium(VII) manganate titration but a redox indicator is added. Potassium dichromate(VI), $K_2Cr_2O_7$, contains dichromate ions, $Cr_2O_7^{2-}$. For example:

$6Fe^{2+}(aq) + Cr_2O_7^{2-}(aq) + 14H^+(aq) \rightarrow 6Fe^{3+}(aq) + 2Cr^{3+}(aq) + 7H_2O(l)$
light green   orange   yellow   deep green

In this titration:

- A redox indicator such as sodium diphenylaminesulphonate is added to the Fe^{2+} solution in the flask. This is because we cannot see an obvious and sudden colour change when a small volume of orange solution is added to a green solution containing Cr^{3+} ions.
- Potassium dichromate(VI) is added gradually from the burette to the acidified solution of iron(II) sulphate in the flask.
- When the colour in the flask changes from greenish to deep purple, the end point has been reached.

Sodium thiosulphate titrations

Sodium thiosulphate, $Na_2S_2O_3$, is useful for determining the concentration of iodine in solution. The redox rection is:

$2Na_2S_2O_3(aq) + I_2(aq) \rightarrow Na_2S_4O_6(aq) + 2NaI(aq)$
colourless   brown   colourless   colourless

In this type of titration, we are often determining the iodine released by another reaction. For example, the reduction of iodate ions by iodide ions in acidic solution:

$IO_3^-(aq) + 5I^-(aq) + 6H+ \rightarrow 3I_2(aq) + 3H_2O(l)$

The titration of the iodine liberated in such reactions gives us a method of determining the amount of oxidising agent, such as IO_3^-, present in a solution.

In sodium thiosulphate titrations:

- Sodium thiosulphate is added gradually from the burette to the (acidified) solution of iodine in the flask.
- When the iodine in the flask has become very pale yellow, add a few drops of starch. This produces an intense blue–black colour. The starch sharpens the end point.
- When just enough sodium thiosulphate has been added to the flask to react with all the iodine, the addition of a further drop of thiosulphate results in the disappearance of the blue colour. A colourless solution is formed.

Exam tips

In potassium manganate(VII) titrations, you are allowed to read the burette from the top of the meniscus rather than the bottom. This is because the colour is so intense that you cannot see the bottom properly. When doing permanganate titrations leave a little time before reading the burette so that the colour on the side of the burette just above the meniscus is minimised.

Key points

- Redox titrations involve oxidising agents such as potassium manganate(VII) or sodium thiosulphate.
- In many redox titrations an indicator is not added because one of the reactants acts as an indicator by giving a specific colour when in excess.
- The colour change in titrations where potassium manganate(VII) is involved is from purple to colourless or colourless to purple.
- Redox titrations involving potassium dichromate(VI) usually need an added redox indicator.
- The amount of iron in iron tablets can be determined by a redox titration with potassium manganate(VII).

Learning outcomes

On completion of this section, you should be able to:

- describe examples of titrimetric analysis in the quantification of substances (vinegar, household cleaners, vitamin C tablets, aspirin, antacids).

Introduction

In Section 8.3, we saw how potassium manganate(VII) can be used to determine the mass of iron(II) sulphate present in iron tablets. Many household products or medicines such as vinegar and aspirin are acidic. Others, such as indigestion (antacid) tablets and some household cleaners are alkaline. Titrimetric analysis can be used to determine the amount of acid or alkali present in these substances.

Determining the acid content of vinegar

The acid present in vinegar is mainly ethanoic acid. The total acid in vinegar can be found by titration of a sample of vinegar with sodium hydroxide of known concentration. Phenolphthalein is used as an indicator as ethanoic acid is a weak acid and sodium hydroxide is a strong base (see *Unit 1 Study Guide*, Section 9.6). To get a suitable titration value, the vinegar is usually diluted by a factor of two.

Did you know?

Good quality vinegar contains at least 5% acid by volume. Vinegar can be made from a great variety of sources (apples, other fruits, rice, sugar cane, palm, etc.) by fermentation. The acetic acid bacteria used in the fermentation process convert ethanol to acetic (ethanoic) acid. Although ethanoic acid is the main acid in vinegar, other natural plant acids may be present in small amounts.

Analysis of aspirin

Aspirin can be hydrolysed to two acids, salicylic acid and ethanoic acid. Hydrolysis and neutralisation happen at the same time if the aspirin is heated with excess sodium hydroxide.

$$CH_3COOC_6H_4COOH + 2NaOH \rightarrow CH_3COONa + HOC_6H_4COONa + H_2O$$

Back titration can be used to quantify the amount of acid present.

The procedure is:

- Put a known mass of aspirin tablets in a flask and boil with excess sodium hydroxide of known concentration and volume for 10 minutes.
- Cool the mixture and titrate the excess sodium hydroxide with $0.05\,mol\,dm^{-3}$ H_2SO_4 using phenol red or phenolphthalein indicator.
- The 'acid content' of the acid is found by: mol NaOH used in the hydrolysis – mol NaOH from titration.

Analysis of antacid tablets

Many antacid tablets contain magnesium hydroxide. The amount of magnesium hydroxide present can be found by:

- crushing the tablet, then reacting it with excess hydrochloric acid of known concentration and volume:

$$Mg(OH)_2(aq) + 2HCl(aq) \rightarrow MgCl_2(aq) + 2H_2O(l)$$

- back titrating the excess acid with sodium hydroxide using screened methyl orange indicator
- the amount of magnesium hydroxide can be found from: mol HCl added to the tablet – mol HCl from titration

Analysis of household cleaners

Sodium chlorate(I) in bleach

Many bleaches contain sodium chlorate(I), NaOCl. This is commonly called sodium hypochlorite. After suitable dilution, the bleach is treated with excess acidified potassium iodide.

$$NaOCl(aq) + 2KI(aq) + H_2SO_4(aq) \rightarrow I_2(aq) + NaCl(aq) + K_2SO_4(aq) + H_2O(l)$$

The iodine liberated is then titrated with standard sodium thiosulphate solution, adding starch indicator as the colour of the iodine fades.

Hydrogen peroxide in household cleaners

Some household cleaners contain hydrogen peroxide. Hydrogen peroxide can be determined by:

- titration with acidified potassium manganate(VII)

$$2MnO_4^-(aq) + 6H^+(aq) + 5H_2O_2(aq) \rightarrow 2Mn^{2+}(aq) + 8H_2O(l) + 5O_2(g)$$

- by adding excess potassium iodide under acidic conditions and titrating the iodine liberated with sodium thiosulphate.

$$2I^-(aq) + H_2O_2(aq) + 2H^+(aq) \rightarrow I_2(aq) + 2H_2O(l)$$

Determining vitamin C

Vitamin C (ascorbic acid) is found in many fruits, especially citrus fruits such as oranges and lemons. It is a good reducing agent. It can be determined using a redox indicator called 2,6-dichlorophenolindophenol (DCPIP). This indicator is blue in colour in its oxidised form. It is colourless in its reduced form, or pink if the conditions are acidic. On addition of vitamin C, DCPIP goes colourless (or pink) as the vitamin C reduces it.

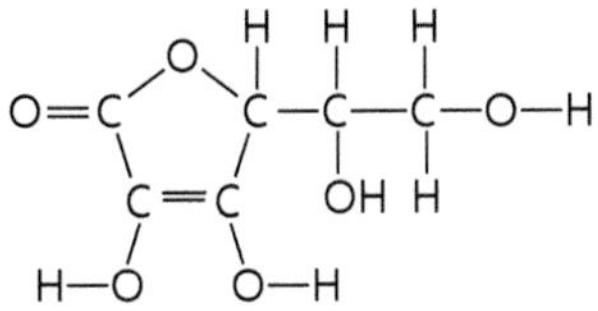

Figure 8.4.1 The structure of vitamin C (ascorbic acid)

The vitamin C content of fruit juice can be determined in the following way:

- Pipette a known volume of fruit juice (suitably diluted) into a titration flask.
- Add 1% DCPIP solution from a burette, drop by drop to the vitamin C solution and shake the flask gently.
- The end point is when the blue colour of the final drop of DCPIP does not fade when added to the solution.

The titration value can be compared with the values obtained by titrating a known concentration of pure vitamin C with the same solution of DCPIP.

> ☑ *Exam tips*
>
> You do not have to know the structure of vitamin C, DCPIP or other redox indicators. You should be aware, however, that redox indicators have two colour forms, one in the reduced state and one in the oxidised state.

> *Key points*
>
> - The acid in vinegar can be determined by titration with sodium hydroxide.
> - The amount of salicylic acid in aspirin tablets can be determined using a back titration.
> - A back titration can be used to determine the amount of magnesium hydroxide present in antacid tablets.
> - The amount of available chlorine in household cleaners can be determined using a sodium thiosulphate titration.
> - Vitamin C can be determined using a redox indicator.

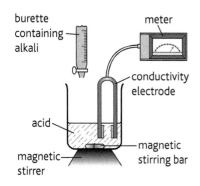

Figure 8.5.1 *Apparatus for a conductimetric titration. The concentration of alkali in the burette is about 20 times the concentration of the acid in the beaker to minimise dilution.*

Conductimetric titrations

Some ions conduct electrical charge better than others. For example, the conductivity of H^+ and OH^- ions is very high compared with that of Cl^- or SO_4^{2-} ions. During a reaction where ions are produced or consumed, there are changes in electrical conductivity. These changes can be measured by **conductimetric titration** using the apparatus shown in Figure 8.5.1.

In an acid–base titration, the base is added in small measured amounts from the burette to the acid in the beaker. After each addition, the meter reading is taken. Typical results for strong and weak acids and bases are shown in Figure 8.5.2.

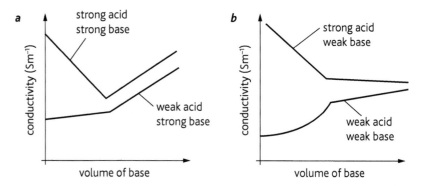

Figure 8.5.2 *Changes in electrical conductivity for titrations involving strong and weak acids and bases; **a** With strong bases; **b** With weak bases*

The end point in these titrations is shown by the 'break-point' in the graph. You can see that you can use this method to find the end point of a weak acid–weak base titration.

Strong acid–strong base: Both acid and base are fully ionised. As the titration proceeds, OH^- ions combine with H^+ ions:

$$H^+(aq) + OH^-(aq) \rightarrow H_2O(l)$$

The conductivity falls as more water is formed. After the end point, there is an excess of OH^- ions and conductivity rises again.

Weak acid–strong base: The acid is only partially ionised but the base is fully ionised:

$$CH_3COOH(aq) + OH^-(aq) \rightleftharpoons CH_3COO^-(aq) + H_2O(l)$$

The conductivity is low to start with because there are few H^+ ions in solution. There is a break in the conductivity curve because OH^- ions (when in excess) are better conductors than CH_3COO^- ions.

Weak base–strong acid: The base is only partially ionised but the acid is fully ionised:

$$H^+(aq) + NH_3(aq) \rightleftharpoons NH_4^+(aq)$$

There is a break in the conductivity curve because H^+ ions (when in excess) are better conductors than NH_4^+.

Potentiometric titrations

Potentiometric titrations involve measuring a change in electrode potential, E, as the titration proceeds. An example is the titration of iron(II) ions with cerium(IV) ions.

$$Fe^{2+}(aq) + Ce^{4+}(aq) \rightarrow Fe^{3+}(aq) + Ce^{3+}(aq)$$

The apparatus is shown in Figure 8.5.3. A standard calomel electrode is used rather than a hydrogen electrode because the latter is bulky and the platinum is easily 'poisoned'.

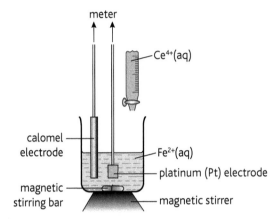

Figure 8.5.3 *Apparatus for a potentiometric titration*

The solution of Ce^{4+} ions is added in small measured amounts from the burette to the Fe^{2+} ions in the beaker. After each addition, the meter reading is taken. At the equivalence point when the Fe^{2+} ions have completely reacted with the Ce^{4+} ions, there is a sharp change in the value of E (Figure 8.5.4).

Potentiometric titrations are useful when coloured solutions such as potassium dichromate or potassium manganate(VI) are used as titrants.

Thermometric titrations

Thermometric titrations are useful when a reaction produces significant enthalpy changes. It can be applied to acid–base reactions, redox reactions or displacement reactions, including precipitation reactions. A solution is added in small measured amounts from the burette to the substance in an insulated beaker with continuous stirring. After each addition, the temperature is taken. Typical results for an exothermic and an endothermic reaction are shown in Figure 8.5.5. You will notice that the equivalence point is shown by a sharp break in the curve.

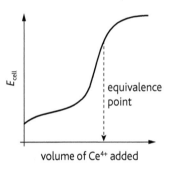

Figure 8.5.4 *Change in electrode potential for the titration of Fe^{2+} ions with Ce^{4+} ions*

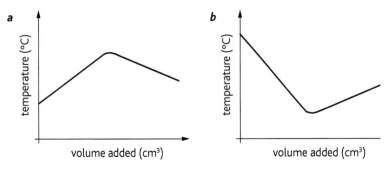

Figure 8.5.5 *Change in temperature during thermometric titrations **a** for an exothermic reaction; **b** for an endothermic reaction*

Key points

- In potentiometric, thermometric and conductimetric titrations, the end point is shown by a sharp break in the line of the relevant graph.

- Conductimetric titrations depend on the relative mobility of the ions present as the reaction proceeds.

- Potentiometric titrations depend on the change in E values as the reaction proceeds.

Learning outcomes

On completion of this section, you should be able to:

- understand the principles on which gravimetric analyses are based
- describe the function of some equipment used in gravimetric analysis (suction flask, suction funnel, silica and sintered-glass crucibles, ovens and furnaces)
- describe how gravimetric analysis is used to determine the moisture content of soils and to find the amount of water in hydrated salts.

Introduction

Gravimetric analysis involves weighing a compound of known composition to determine the amount of one of the substances present. The main steps are:

- preliminary treatment, e.g. dissolving or pH adjustment
- precipitation
- filtration
- washing the precipitate
- drying or ignition of the precipitate
- weighing the dried precipitate (to three or four decimal places)
- calculation of the amount of the element to be determined.

Did you know?

Theodore Richards was the first American to win a Nobel Prize for Chemistry (1914). He developed many of the techniques of gravimetric analysis. He used these techniques to determine accurate atomic masses of about 25 elements.

Precipitating and filtering a sample

A suction flask and funnel is used to filter and wash precipitates. Figure 8.6.1 shows a sintered-glass (ground glass) crucible and a suction funnel used for filtration.

All the solid must be transferred to the funnel by washing out the container containing the precipitate into the funnel:

- The pump is turned on and the liquid to be filtered is directed into the sintered-glass crucible down a glass rod.
- Any remaining solid is transferred to the crucible using a gentle stream of water until no solid remains in the beaker or glass rod.

The suction (Buchner) funnel is not used in accurate quantitative work but is useful for filtering salts which have been purified by crystallisation.

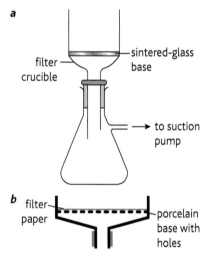

Figure 8.6.1 *Two pieces of apparatus for filtration; **a** A sintered-glass crucible; **b** A suction funnel (Buchner funnel)*

Washing a sample

When washing a precipitate care must be taken to prevent any redissolving.

- The precipitate is washed for no longer than necessary.
- Small amounts of water or other solvent are used.
- Washing with a solution containing an ion which is common to one in the precipitate makes it less likely that any precipitate will redissolve (see *Unit 1 Study Guide*, Section 8.7).

Drying

The precipitate in the porous sintered-glass crucible can be oven-dried if necessary so that the precipitate does not have to be transferred to a separate container. The mass of precipitate can then be found from (mass of crucible + ppt – mass of crucible alone).

- Surface water can be removed from a precipitate by drying for 1–2 hrs at 110°C.
- All water is removed when heated at a higher temperature in an oven.
- When a precipitate is heated at a high temperature in a furnace to red heat we say that it is 'ignited'.
- Drying or 'ignition' of the precipitate is carried out several times until a constant mass is obtained.
- Compounds which are likely to be decomposed by heating, e.g. hydrated salts should be air-dried. The damp solid is left to dry for a day in a place free of dust.
- A desiccator containing phosphorus(v) oxide, silica gel or soda lime can be used to dry many solids but should not be used for hydrated salts, since some water of crystallisation may be lost.

Finding the loss of mass on heating

This method can be used to determine the moisture content of soils and to find the amount of water in hydrated salts.

Water of crystallisation of barium chloride

$$BaCl_2 \cdot nH_2O(s) \rightarrow \quad BaCl_2(s) \quad + \quad nH_2O(g)$$
original mass   mass of residue   mass lost

- Weigh a clean empty crucible (m_1).
- Half fill the empty crucible with $BaCl_2 \cdot nH_2O$ and reweigh (m_2).
- Heat gently at first then more strongly to red heat.
- Let the crucible cool completely then reweigh.
- Reheat as many times as necessary until constant mass is obtained (m_3).

The loss in mass $(m_2 - m_3)$ and the residual mass $(m_3 - m_1)$ can be used to calculate the number of moles of water of crystallisation per mole of $BaCl_2 \cdot nH_2O$. For the calculation see Section 8.7.

Determining the moisture content of soils

The process is similar to that above. An accurately-weighed sample of soil is heated at about 110°C to constant mass. A low temperature is used to prevent organic material from being burnt. The loss of mass is due to the loss of water. The percentage (%) of water by mass in the soil can then be calculated.

Key points

- Gravimetric analysis involves weighing a compound of known composition to determine the amount of one of the substances present.
- The main steps in gravimetric analysis are precipitation, filtration, washing, drying and weighing.
- Loss of material in washing precipitates is reduced by using a wash material with a common ion.
- Drying of material should be carried out to a constant weight and without decomposition.

☑ *Exam tips*

If asked about how to dry a substance you need to know whether the substance:

1 decomposes readily

2 reacts with any drying agent used in a dessicator

3 readily gains water from the air or readily loses water to the air.

Determining the number of moles of water in a hydrated salt

In the last section we introduced an experiment to calculate the number of moles of water in hydrated barium chloride. By weighing a sample of barium chloride before and after heating, the mass of water lost can be found.

$$BaCl_2 \cdot nH_2O(s) \rightarrow \quad BaCl_2(s) \quad + \quad nH_2O(g)$$
$$\text{original mass} \quad \text{mass of residue} \quad \text{mass lost}$$

Worked example 1

When 0.611 g of hydrated barium chloride is heated to constant mass: 0.521 g of residue are formed. Deduce the formula of hydrated barium chloride. (A_r values: Ba = 137.3, Cl = 35.5, O = 16).

Step 1: Calculate the loss of mass of water: 0.611 − 0.521 = 0.090 g

Step 2: Calculate the number of moles of water = $\dfrac{0.090}{18}$ = 5 × 10⁻³ mol

Step 3: Calculate the moles of residue ($BaCl_2$)

$$= \frac{0.521}{208.3} = 2.5 \times 10^{-3}\,\text{mol}$$

Step 4: Calculate the mole ratio of water to $BaCl_2$

$$BaCl_2 : H_2O = 2.5 \times 10^{-3} : 5 \times 10^{-3} = 1:2 \text{ ratio}$$

So formula is $BaCl_2 \cdot 2H_2O$.

Determination of chlorides

Chlorides can be determined by:

- weighing a sample of the solid metal chloride
- dissolving the sample in water
- adding nitric acid then excess silver nitrate to the chloride solution. The chloride precipitates as silver chloride:

$$Cl^-(aq) + AgNO_3(aq) \rightarrow AgCl(s) + NO_3^-(aq)$$

- collecting the chloride formed by filtration
- washing the chloride with distilled water
- drying the chloride
- weighing the mass of silver chloride formed.

Note: The experiment should be carried out in a darkened room because silver chloride is sensitive to light.

Worked example 2

A 0.497 g sample of a chloride of a Group I metal, Z, is dissolved in water. Excess acidified silver nitrate is added to the solution. The resulting precipitate is filtered and dried to constant mass. The mass of silver chloride formed is 0.957 g. Deduce which metal is present in the original chloride. (AgCl = 143.5 g mol⁻¹; A_r chlorine = 35.5).

Step 1: Calculate the moles of silver chloride:

$$= \frac{0.957}{143.5} = 6.67 \times 10^{-3}\,mol$$

Step 2: Calculate the mass of Cl⁻ ions in AgCl:

$$= 6.67 \times 10^{-3} \times 35.5 = 0.237\,g$$

Step 3: Calculate the mass of metal in the metal chloride:

$$0.497 - 0.237 = 0.260\,g \text{ of } Z$$

Step 4: Calculate the moles of metal present and so atomic mass:

Since Z is a Group I metal, 1 mol of Z forms 1 mol of chloride ions:

$$ZCl(aq) + AgNO_3(aq) \rightarrow AgCl(s) + ZNO_3(aq)$$

So: $mol\ Z = \dfrac{mass\ of\ Z}{atomic\ mass\ of\ Z}$

$$atomic\ mass\ of\ Z = \frac{mass\ of\ Z}{moles\ Z} = \frac{0.260}{6.67 \times 10^{-3}} = 38.9$$

Potassium is the Group I metal which has an atomic mass of 39.

Worked example 3

Deduce the formula of magnesium chloride from the following information:

0.635 g of magnesium chloride, $MgCl_x$ reacts with excess silver nitrate. The mass of silver chloride formed is 1.914 g. ($AgCl = 143.5\,g\,mol^{-1}$; A_r chlorine = 35.5, A_r magnesium = 24).

Step 1: Moles of silver chloride: $= \dfrac{1.914}{143.5} = 0.0133\,mol$

Step 2: Mass of Cl⁻ ions in AgCl: $= 0.0133 \times 35.5 = 0.472\,g$

Step 3: Mass of magnesium in the magnesium chloride:

$$0.635 - 0.472 = 0.163\,g$$

Step 4: Moles and mole ratio:

$$mol\ Mg = \frac{0.163}{24} = 6.79 \times 10^{-3}\,mol$$

So mole ratio $= 6.79 \times 10^{-3}\,mol\ Mg: 1.33 \times 10^{-2}\,mol\ Cl$

which is 1 mol Mg: 2 mol Cl

So formula is $MgCl_2$.

Gravimetric analysis in quality control

Gravimetric analysis can be used to:

- determine the amount of elements such as phosphorus in fertilisers (by conversion to insoluble magnesium ammonium phosphate)
- determine sulphur dioxide in the air and in wine or fruit drinks (by conversion to barium sulphate)
- determine the chloride ions present in our water supply (by conversion to silver chloride).

Key points

- Calculations based on gravimetric analysis can be used to find the molar composition of particular compounds.
- Calculations based on gravimetric analysis can be used to calculate the number of water in one mole of a hydrated salt.
- Gravimetric analysis can be used in quality control to determine the % water in soil and in foods and to determine the amount of particular elements in soil and foodstuffs.

9 Spectroscopic methods

9.1 Electromagnetic radiation

Did you know?

About 100 years ago sources emitting gamma-rays (γ-rays) were put under the pillows of some people, under the mistaken belief that it would help them sleep. Now we know that these rays are very dangerous.

The electromagnetic spectrum

Electromagnetic radiation consists of waves that have electrical and magnetic components vibrating in particular directions, e.g. light waves. The various types of electromagnetic radiation are shown in Figure 9.1.1.

Each of these types of radiation has specific ranges of wavelengths and **frequencies**. For example, light waves have a frequency of $4.5 \times 10^{14} - 7.5 \times 10^{14}$ Hz. All these types of electromagnetic radiation travel at the same speed in air, 3.0×10^8 (or $300\,000\,000$) m s^{-1}.

Speed, frequency and wavelength

The speed of electromagnetic radiation is related to frequency and wavelength by the equation:

$$c = f\lambda$$

where: c is speed in m s^{-1}

f is frequency (number of wave crests per second) in hertz, Hz. ($1\,\text{Hz} = 1\,\text{s}^{-1}$)

λ is the wavelength in m

Note: The symbol ν is often used for frequency when electrons are being considered.

The smaller the wavelength, the greater is the frequency (the more waves per second). The greater the frequency, the greater is the amount of energy transferred. So gamma-rays (γ-rays) and X-rays carry a huge amount of energy. Because of this they are very dangerous. They can easily penetrate the skin and damage cells in the body. Even ultraviolet

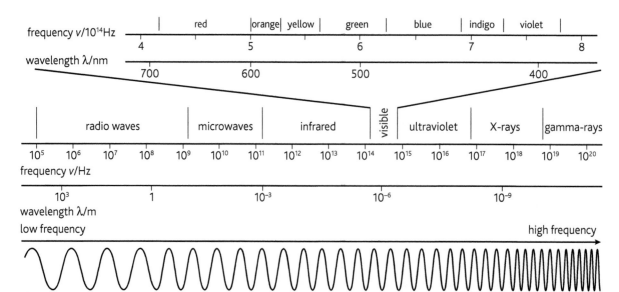

Figure 9.1.1 The electromagnetic spectrum

rays (UV rays) have enough energy to cause harmful burns, skin cancer and damage to the eyes if we are exposed to them for long enough.

Energy quanta

In *Unit 1 Study Guide*, Section 1.3, we learnt that electrons in atoms only have certain fixed values of energy. These values are called **quanta**. Figure 9.1.3 shows the movement of an electron between energy levels in an atom.

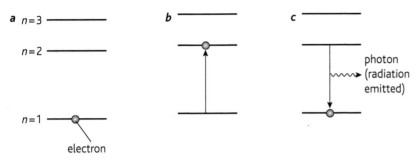

Figure 9.1.3 *Movement of an electron between energy levels;* **a** *Atom in the ground state;* **b** *Excited electron;* **c** *Electron falling back to ground state*

- When a quantum of energy is absorbed by an electron in the ground state atom the electron is excited to a higher energy level.
- When an electron falls back to the ground state again, it gives out a quantum of energy as radiation. We can also think of this energy as a particle called a photon.
- The energy difference involved is given by the equation:

$$\underset{\substack{\uparrow \\ \text{energy (J)}}}{\Delta E} = h\underset{\substack{\downarrow \\ \text{frequency of} \\ \text{radiation (Hz)}}}{v}$$

Planck constant
$(6.63 \times 10^{-34} \, \text{J Hz}^{-1})$

Since: frequency $= \dfrac{\text{speed of light}}{\text{wavelength}}$ or $v = \dfrac{c}{\lambda}$

we can also write: $\Delta E = \dfrac{hc}{\lambda}$

We can use this equation to calculate the energy emitted when radiation of a particular wavelength or frequency is emitted from a previously excited atom. If the frequency of radiation is measured at the convergence limit, we can calculate the ionisation energy of the atom (see *Unit 1 Study Guide*, Sections 1.3 and 1.4).

Worked example

Calculate the energy of an electron transition which emits radiation of frequency 1.01×10^{12} Hz.

Planck constant $= 6.63 \times 10^{-34}$ J s.

Substituting into the equation $\Delta E = hv$

$$6.63 \times 10^{-34} \times 1.01 \times 10^{12} = 6.70 \times 10^{-22} \, \text{J}$$

If we are asked for the value per mole of electrons, we multiply the value by the Avogadro number, 6.02×10^{23}.

$$6.70 \times 10^{-22} \times 6.02 \times 10^{23} = 6.02 \times 10^{23} = 403 \, \text{J mol}^{-1}$$

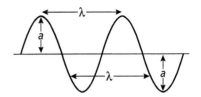

Figure 9.1.2 *Wavelength (λ) and amplitude (a). Wavelength is the distance between any two similar points on the wave.*

Did you know?

Although X-rays are dangerous, they are still used in small 'doses' to 'see' the bones in the body and for sterilising hospital equipment.

Did you know?

In 1906 Einstein suggested that light was packets of energy called photons. In 1925 Louis de Broglie suggested that the energy of a photon is related to the frequency of radiation and to its mass. Nowadays we can regard electrons as having properties of waves as well as particles.

Key points

- Electromagnetic radiation can be regarded as waves that have a characteristic frequency and wavelength.
- Speed = frequency $\times$ wavelength.
- The spectrum of electromagnetic radiation ranges from radio waves (10^7 Hz) to gamma-rays (10^{19} Hz).
- Light waves have a frequency of $4.5 \times 10^{14} - 7.5 \times 10^{14}$ Hz.
- Energy levels in atoms are quantized – they can only have certain energy values.
- The energy associated with a photon is given by $E = hv$, where c is the speed of light and h is Planck constant.

Did you know?

Beer–Lambert's law is a combination of two laws: Beer's law which refers to the effect of the absorbed light on concentration and Lambert's law which refers to the absorbance of light by a pure liquid.

Introduction

In *Unit 1 Study Guide*, Sections 7.1 and 13.3 we saw how a colorimeter is used to determine the concentration of coloured molecules or ions in solution. The filter is chosen to select a band of wavelengths that are absorbed by the solution. A calibration curve for a particular filter and colorimeter is needed (Section 7.3). A UV-visible absorption spectrometer measures the concentration of specific substances using a narrow range of wavelengths in the UV and visible regions. Both a colorimeter and UV-visible spectrometer depend on Beer–Lambert's law.

Beer–Lambert's law

Transmittance refers to light passing through a solution. **Absorbance** refers to light absorbed by a solution. If all the light passes through a solution the absorbance is 0% and the transmittance 100%. If no light passes through a solution, the absorbance is 100% and the transmittance is 0%. Absorbance (A) is related to transmittance (T) by the equation:

$$A = -\log_{10} T$$

Beer–Lambert's law states that:

- the amount of light absorbed is proportional to the concentration of the solution
- the amount of light absorbed is proportional to the distance it travels through the solution (the path length, l).

In symbols:

proportionality constant

absorbance $— A = \varepsilon l c$ —concentration of solution

distance light travels through solution

The constant, ε, is called the molar absorptivity. It gives a value of the absorbance, A, when the light travels for 1 cm through a solution of $1\ mol\,dm^{-3}$. It has units of $mol^{-1}\ dm^3\,cm^{-1}$ but is commonly quoted without units.

The absorbance is also given by the relationship:

$$\log \frac{I_o}{I} \propto c$$

- I_o is intensity of light transmitted through a colorimeter cell containing pure solvent.
- I is the intensity of the light transmitted through the cell containing the solution under test.

If the meter readings on the colorimenter are proportional to the light intensity on the detector, then:

$$\log \frac{\text{transmittance in cell with pure solvent}}{\text{transmittance in cell containing test solution}} = \text{constant} \times c$$

We can use Beer–Lambert's law to do simple calculations:

Worked example 1

When radiation of wavelength 200 nm is passed through solution X in a cell of 1.5 cm path length, its absorption measured on a spectrometer is 0.65. Solution X has a molar absorptivity of 80. Calculate the concentration of the solution.

- Rearrange Beer–Lambert's law in the form $A = \varepsilon lc$ so that c is the subject:

$$c = \frac{A}{\varepsilon l}$$

- Substitute the values:

$$c = \frac{0.65}{80 \times 1.5} = 5.4 \times 10^{-3}\,\text{mol dm}^{-3}$$

Worked example 2

A solution containing $0.025\,\text{mol dm}^{-3}$ copper(II) sulphate (solution A) is placed in a spectrometer cell of path length 1 cm. The absorbance of this solution is 0.48. Another solution of copper sulphate (solution B) is placed in the same cell under the same conditions. The absorbance is 0.18. Calculate (i) the concentration of solution B (ii) the molar absorptivity of copper(II) sulphate.

(i) Since absorbance is proportional to concentration if all other factors are constant

absorbance of $0.48 \rightarrow 0.025\,\text{mol dm}^{-3}$ copper(II) sulphate.

So absorbance of $0.18 \rightarrow 0.025 \times \dfrac{0.18}{0.48} = 0.0094$ (to 2 s.f.)

(ii)

- Rearrange Beer–Lambert's law in the form $A = \varepsilon lc$ so that ε is the subject:

$$\varepsilon = \frac{A}{lc}$$

- Substitute the values:

$$\varepsilon = \frac{0.48}{1 \times 0.025} = 19.2$$

Deviation from Beer–Lambert's law

The relationship:

$$\log \frac{\text{transmittance in cell with pure solvent}}{\text{transmittance in cell containing test solution}} = \text{constant} \times c$$

is not obeyed when high concentrations of solutions are used, especially coloured solutions. The absorbance increases less rapidly as the concentration increases. For this reason we have to plot a calibration curve in order to calculate the concentration of a particular solution accurately (see Section 7.3).

Beer–Lambert's law may seem quite complicated at first sight but all you have to remember is:

the absorbance is proportional to the concentration of solution and the absorbance is proportional to the path length of the cell.

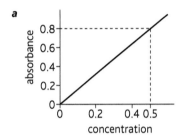

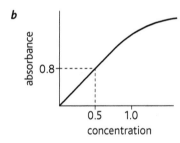

Figure 9.2.1 a At low concentrations a graph of absorbance against concentration shows proportionality. It follows Beer–Lambert's law. **b** At higher concentrations, Beer–Lambert's law is not obeyed.

Key points

- Transmittance refers to light passing through a solution. Absorbance refers to light absorbed by a solution.

- Beer–Lambert's law states that the amount of light absorbed is proportional to the concentration of the solution and to the distance it travels through the solution.

- Beer–Lambert's law can be expressed as: $A = \varepsilon lc$, where A is absorbance, c is the concentration of solution, l is the path length and ε is a constant (molar absorptivity).

- At high concentrations of solute, Beer–Lambert's law is no longer obeyed.

On completion of this section, you should be able to:

- understand the use of UV spectra
- describe the steps in analysing samples by UV spectroscopy
- understand about sensitivity and detection limits in UV spectroscopy
- describe the steps in analysing samples by visible spectroscopy
- understand the use of visible spectroscopy
- understand the use of complexing reagents to form coloured compounds (sensitivity and detection limits).

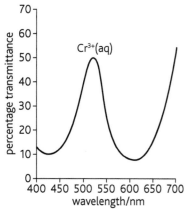

Figure 9.3.1 *Absorption spectrum of a solution containing Cr^{3+} ions*

The UV-visible absorption spectrometer

A single beam spectrometer is similar to a colorimeter (see Section 7.3 for diagram) but it uses a diffraction grating to select wavelengths in the **UV** or **visible** region rather than a filter. The procedure is:

- Set the wavelength of light required.
- Place pure solvent in the cell (water, ethanol or other organic solvent).
- Adjust the meter reading to 0 absorbance or 100% transmittance.
- Put the sample in another identical cell and place this in the path of the light.
- Record the meter reading (absorbance or transmittance).
- Repeat at other selected wavelengths.
- A calibration curve using standard solutions can be used to relate the absorbance to the concentration of the substance present (see Section 7.3 for details).

A single beam spectrometer can be used to identify coloured ions in solution (Figure 9.3.1) but it cannot be used to distinguish between organic molecules satisfactorily. The wavelength is usually measured in nanometres, nm ($1\,nm = 10^{-9}\,m$).

Using complexing reagents

Some ions may not absorb light very well in the visible or UV regions. They may however be converted to more highly coloured ions by forming complex ions with particular ligands (see *Unit 1 Study Guide*,1 Section 13.3). These ions may have better absorption characteristics. They are more sensitive to the absorption of light. They may also shift the wavelength of maximum absorption. For example:

- Fe^{2+} ions in solution are light green in colour. If present in low concentrations they do not absorb visible radiation very well. When reacted with the ligand 1,10-phenanthroline, however, a deep orange–red complex ion is formed. At the appropriate wavelength this complex absorbs radiation to a far greater extent than the Fe^{2+} ions alone. This maximises the precision of the measurements.
- The colour of dilute copper(II) sulphate is due to a water–Cu^{2+} ion complex. This is very light blue in colour. Adding ammonia makes a deep blue complex and changes the wavelength of maximum absorption.

High resolution UV-visible spectroscopy

High resolution UV-visible spectrometers have two separate light sources, one for the UV and one for the visible regions. The wavelength range used is about 200–800 nm. A simplified diagram of this spectrometer is shown in Figure 9.3.2.

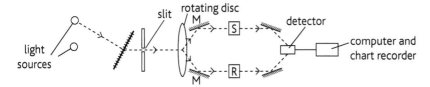

Figure 9.3.2 *A double beam UV-visible spectrophotometer. M = mirror. S = sample cell. R = reference cell.*

- The diffraction grating rotates to allow light from the whole range of the UV or visible regions to be produced.
- The rotating disc divides up the beam so that it alternates between the cell containing the test solution and a reference cell containing the solvent used in preparing this solution.
- The detector compares the values of the sample and reference cells and converted to percentage (%) transmittance then to molar absorptivity.
- Cells and lenses made of quartz are used because glass absorbs radiation in the ultraviolet region.

A typical UV-absorption spectrum is shown in Figure 9.3.3. Different compounds can be identified by their typical absorption peaks. For butanone there are two peaks, one at a wavelength of 190 nm and another at 275 nm. We can use such spectra to distinguish between different types of organic compound.

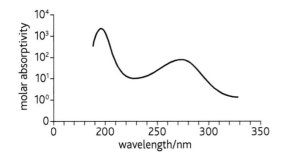

Figure 9.3.3 *The UV-absorption spectrum of butanone*

Limitations of UV-visible spectroscopy

A UV or visible region spectrum is not enough to identify a substance with absolute certainty because:

- The solvents used may absorb UV radiation significantly.
- The polarity of the solvent and pH can affect the UV absorption spectrum.
- The temperature and high electrolyte concentration may interfere with the spectrum.
- The method is limited to either:
  - coloured compounds (in visible spectroscopy) or
  - organic compounds with conjugated double bonds such as alkenes with alternating double and single bonds and carbonyl compounds (in UV spectroscopy).
- The width of the spectrometer slit and other variables associated with the spectrometer also affect the spectrum.

Key points

- Some molecules absorb radiation in the UV or visible region. This gives rise to characteristic spectra.

- Visible spectroscopy is limited to coloured compounds.

- UV spectroscopy is used for organic molecules with conjugated double bonds or carbonyl compounds.

- Samples are analysed by UV-visible spectroscopy by passing radiation through a quartz cell and detecting the radiation transmitted compared with a reference cell.

- The sensitivity and detection limits in UV spectroscopy are low compared with other spectroscopic methods.

- Complexing reagents can be added to many ions to increase the sensitivity and detection limits.

Learning outcomes

On completion of this section, you should be able to:

- explain the origin of UV spectra

- explain why some molecules absorb radiation in the UV and visible regions and others do not

- list examples of the use of ultraviolet spectra in the quantization of substances (iron tablets, glucose and urea in blood, cyanide in water).

Did you know?

When atomic orbitals overlap, two molecular orbitals are formed. One of the molecular orbitals has lower energy than the atomic orbitals. This is called a bonding orbital. The other has a higher energy than the atomic orbitals. This is called an antibonding orbital (*). Both σ and π bonds have these orbitals. There are also non-bonding orbitals (n). These usually contain lone pairs of electrons.

Figure 9.4.1 *The relative energies of different orbitals*

When an electron is excited by ultraviolet light, the electron generally moves to the antibonding orbital.

Which molecules absorb UV/visible radiation?

The absorption of light in transition element ions

- The d orbitals in an isolated transition element ion are described as degenerate. They all have the same average energy.

- In the presence of ligands, the orbitals split into two groups (see *Unit 1 Study Guide*, Section 13.3).

- When energy in the visible region is absorbed an electron moves from a d orbital of lower energy to a d orbital of higher energy. Light is absorbed in the visible region of the spectrum.

- The wavelength of the light absorbed depends on the energy difference between the split d levels.

The absorption of radiation in organic compounds

Organic compounds which absorb radiation in the ultraviolet (or visible) regions usually exhibit a resonance structure. Two examples are:

- conjugated bonds (alternating double and single bonds) in dienes, e.g. buta-1,3-diene, $CH_2 = CH—CH = CH_2$

- the $C = O$ group in aldehydes and ketones, e.g. butanone

$$CH_3CH_2\overset{\overset{\displaystyle O}{\|}}{C}CH_3$$

When UV radiation passes through these compounds, energy from the radiation is used to move an electron in the outer electron shell from a lower energy level to a higher energy level. These energy levels are often associated with orbitals that contain pi (π) electrons.

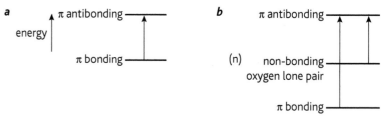

Figure 9.4.2 *The absorption of energy in the ultraviolet region moves an electron from a lower energy level to a higher energy level. The orbitals involved are shown for **a** a diene, **b** a ketone.*

The energy involved in moving electrons from $π \rightarrow π^*$, $n \rightarrow π^*$ or $n \rightarrow σ^*$ energy levels is sufficient to cause absorption in the UV region. Movements from the σ bonding energy level to the $π^*$ or $σ^*$ levels usually require too much energy to cause absorption in the UV region. For buta-1,3-diene, the only electron movements which take place within the ultraviolet range are from pi bonding (π) to pi antibonding ($π^*$) orbitals. Butanone can absorb radiation at two different wavelengths. The peaks result from two different movements of electrons in the $C = O$ bond ($n \rightarrow π^*$ and $π \rightarrow π^*$).

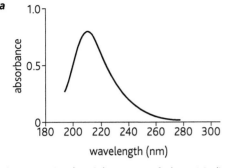
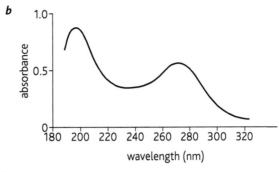

Figure 9.4.3 *Ultraviolet spectra of **a** buta-1,3-diene and **b** butanone*

For a similar group of molecules, e.g. alkenes with one, two or three double bonds, as the delocalisation increases, the peak of maximum absorption moves to longer wavelengths. For example:

$$CH_2=CH_2 \qquad CH_2=CH—CH=CH_2 \qquad CH_2=CH-CH=CH—CH=CH_2$$

peak: 171 nm   217 nm   258 nm

─────────────────── increased delocalisation ───────────────────→

Use of ultraviolet spectra

Although UV and visible spectra can be used to identify particular compounds, the spectra are not as readily interpreted as infrared spectra or mass spectra. Their main use is for determining the amount of a particular substance present in solution by measuring the absorbance at a particular wavelength. An appropriate calibration curve is usually used to relate concentration to absorbance.

- *Iron in iron tablets:* When reacted with the ligand 1,10-phenanthroline, a deep orange-red complex ion is formed. At about 510 nm this complex absorbs radiation well, so can be used for quantifying iron.

- *Glucose in blood:* Glucose is colourless so cannot be quantified by visible spectroscopy. If we react glucose with excess Benedict's solution (blue in colour) an insoluble precipitate of copper(I) oxide is formed. The blue solution becomes less intense in colour. By measuring the colour intensity of the blue solution using a visible spectrometer, we can calculate the glucose concentration. Glucose concentration can also be measured by reacting the glucose with glucose oxidase and other enzymes. The solution is then analysed with a UV-visible spectrometer at a suitable wavelength, e.g. 340 nm.

- *Urea in blood:* This can be analysed by adding zinc sulphate and Ehrlich's reagent. The product formed absorbs radiation at 435 nm. Urea can also be quantified by reacting it with specific enzymes and measuring the absorbance at 340 nm.

- *Cyanide in water:* The cyanide in the water is converted to cyanogen bromide, CNBr, by treatment with bromine water. On addition of *p*-phenylenediamine a red dye is formed, which absorbs radiation at 530 nm.

Did you know?

The cassava plant, which is grown as a root crop in many parts of the world, contains very small amounts of cyanide in its tuberous root. If you do not cook the tubers properly, you risk being very ill through cyanide poisoning.

Key points

- Organic compounds, which absorb radiation in the ultraviolet (or visible) regions, usually have a resonance structure.

- The energy involved in moving electrons from $\pi \to \pi^*$, $n \to \pi^*$ or $n \to \sigma^*$ energy levels causes absorption in the UV region.

- Spectroscopy in the UV-visible region is used to determine the amount of a particular substance present in solution by measuring the absorbance at a particular wavelength.

- Iron tablets, glucose and urea in the blood and cyanide in water can be determined by adding specific reagents or enzymes. The resulting species absorb radiation in the UV-visible region.

On completion of this section, you should be able to:

- explain the origin of absorption of infrared (IR) radiation by molecules
- describe the basic steps involved in analysing samples by IR spectroscopy (referring to preparation of solids)
- describe the limitations of IR spectroscopy.

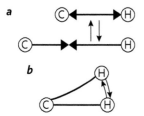

Figure 9.5.1 *Vibrations in a C—H bond; a Stretching vibration; b Bending vibration*

Why do molecules absorb infrared radiation?

In a covalent molecule such as methane the electron clouds bonding the C and H atoms allow the nuclei to vibrate in two ways: **stretching** and **bending**.

- Covalent bonds have a natural frequency of vibration.
- A molecule absorbs **infrared (IR) radiation** whose frequency is the same as the natural vibration of the bonds in the molecule. The energy associated with the vibrations is quantized (see Section 9.1).
- The energy absorbed increases the amplitude of the vibration of the bonds.
- Absorption of IR radiation only happens when:
  - there is some type of charge separation within the molecule (the molecule is polar and hence has a dipole)

  and

  - the vibration results in a change in the dipole moment of the molecule (see *Unit 1 Study Guide*, Section 2.5). So molecules such as H_2 and Cl_2 do not absorb in the infrared region but HBr will.

☑ *Exam tips*

You may find it useful to think of vibrations in bonds rather like springs attached to a pair of atoms. You can stretch and bend these bonds in several ways by supplying energy from your fingers.

Characteristics of an infrared spectrum

The energy absorbed as a result of molecular vibrations depends on the masses of the atoms and the bond strength.

- Each type of bond absorbs IR radiation of a specific frequency.
- Different types of vibration in particular bonds give rise to absorptions in particular regions of the spectrum, e.g. the absorption due to C—H stretching vibrations are at higher frequencies than the absorption due to C—H bending vibrations.
- The IR spectrum shows the percentage (%) of radiation transmitted. It appears as a series of dips (peaks) where particular bonds have absorbed radiation.
- The position of the peaks is given by the **wavenumber** measured in cm^{-1}.

$$\text{wavenumber} = \frac{\text{frequency in hertz}}{\text{speed of light in } cm\,s^{-1}}$$

So the wavenumber corresponding to a frequency of 9×10^{13} Hz is:

$$9 \times 10^{13} / 3 \times 10^{10} = 3000 \, cm^{-1}$$

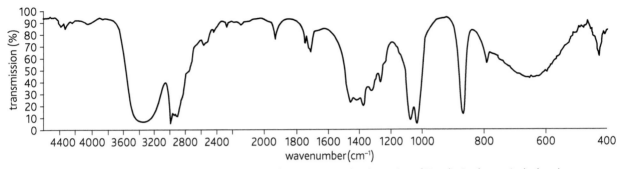

Figure 9.5.2 *A typical infrared spectrum. Each of the main dips represents the absorption of IR radiation by particular bonds.*

The infrared spectrometer

The simplified structure of an infrared spectrometer is shown in Figure 9.5.3.

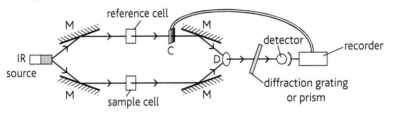

Figure 9.5.3 *Simplified diagram of an infrared spectrometer. M = mirrors. C = Comb. D = rotating disc.*

- A beam of IR radiation is produced from a ceramic rod heated to 1500°C.
- The radiation is passed through the sample cell and reference cell.
- The level of IR radiation of the beam passing through the sample is compared with that coming through the reference cell. The comb is part of the mechanism which helps this comparison.
- The diffraction grating or prism is rotated so that different wavelengths of IR radiation are brought to the detector.
- The recorder plots a graph of % **transmission** against wavenumber.
- The diffraction grating, prism, mirrors and cells cannot be made from glass because glass absorbs IR radiation. The cells are often made from potassium bromide or calcium fluoride. These substances do not absorb IR radiation.

Preparing the sample for infrared spectroscopy

Liquid samples: The sample for analysis is placed as a thin film between two discs of sodium chloride.

Solid samples: The sample for analysis is finely powdered and mixed with potassium bromide or sodium chloride (which do not absorb IR radiation). This mixture (mull) is then crushed to form a disc. Alternatively, the sample for analysis can be powdered and placed between two discs of sodium chloride.

Limitations of infrared spectroscopy

- It cannot be used to identify substances that are non-polar.
- It cannot be used to identify substances that are electrolytes or have ionic components.
- It provides information about the types of groups present, including functional groups, but not always about the structure of the molecule as a whole.

Did you know?

The more complex a molecule, the more ways there are in which a molecules can vibrate. In water there are three ways in which molecules can vibrate but in propanone there are 24 ways. For a linear molecule containing n atoms there are $3n - 5$ possible ways of vibrating.

Key points

- Infrared (IR) radiation is absorbed by molecules when the radiation is the same as the natural frequency of vibration of the molecules.
- The two main types of vibration in molecules are bending and stretching.
- Samples of solids for IR spectroscopy are made by forming them into a disc with KBr or NaCl.
- In an IR spectrometer, the radiation absorbed by the sample is compared with a reference sample.
- There are some limitations to the use of IR spectroscopy, e.g. molecules such as Cl_2 do not absorb IR radiation.

Learning outcomes

On completion of this section, you should be able to:

- deduce chemical groups including functional groups from information from IR spectra

- identify OH, NH_2 $C=O$, $C=C$, COOH and $CONH_2$ groups from IR spectra

- give examples of the use of IR spectra in monitoring air pollutants such as CO_2 and SO_2.

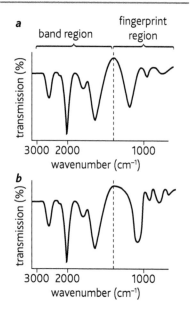

a

band region / fingerprint region

b

Figure 9.6.1 *Simplified infrared spectra of a propanone and b butanone*

The band region and fingerprint region

Particular groups such as C—H, O—H and C=O absorb radiation with wavenumbers in the region of $1300–3000\,cm^{-1}$. Specific peaks indicate the presence of these groups in the molecule. We call this the **band region** of the spectrum. In this region compounds in the same homologous series have almost identical spectra. Peaks in the $600–1300\,cm^{-1}$ wavenumber region of the spectrum tell us about the structure of the whole molecule. We call this the **fingerprint region**. This region can be used to distinguish between molecules with same functional group, e.g. propanone and butanone (Figure 9.6.1).

Identifying specific groups

When identifying groups from IR spectra, we match the peaks (dips) with known values for these groups. Some typical wavenumbers are given in the table. These are due to stretching vibrations; their values may vary according to the types of atoms surrounding them.

Group	alcohol O—H	amine N—H	aldehyde/ ketone C=O	alkene C=C	carboxylic acid O—H
Wavenumber / cm⁻¹	3580– 3650	3350– 3500	1680– 1750	1610– 1680	2500– 3000

The wavenumber of the C=O group may also vary according to its environment:

C=O in carboxylic acid 1700–1725   C=O in amide 1630–1700

C=O in aldehyde 1720–1740   C=O in ketone 1680–1700

C=O in ester 1730–1750

Another useful value is C—O in alcohols, ethers and esters = 1000–1300

Alcohols and other compounds which are highly hydrogen-bonded show a very broad peak between 3230 and $3550\,cm^{-1}$.

Example 1

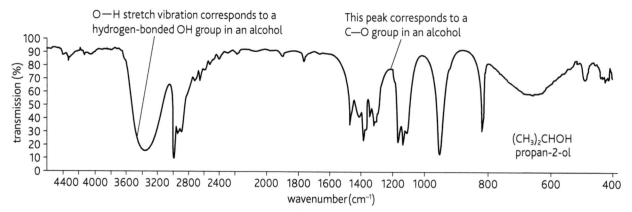

Figure 9.6.2 *Infrared spectrum of propan-2-ol*

Example 2

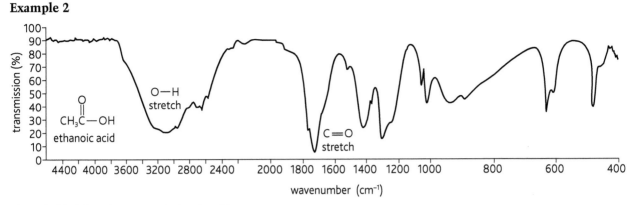

Figure 9.6.3 *Infrared spectrum of ethanoic acid*

We can identify a very wide O—H stretching vibration at 2800–3300 cm⁻¹. This corresponds to a hydrogen-bonded O—H group in a carboxylic acid. There is also a peak at about 1730 cm⁻¹ which corresponds to a C=O group in a carboxylic acid.

Example 3

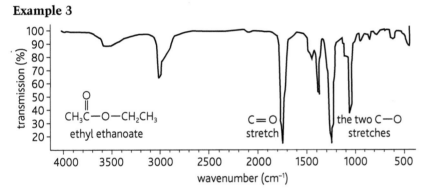

Figure 9.6.4 *Infrared spectrum of ethyl ethanoate*

We can identify a C=O stretching vibration at 1750 cm⁻¹ due to the C=O group in an ester. There are also two peaks at about 1050 cm⁻¹ and 1250 cm⁻¹ which correspond to the two C—O groups in the structure.

Infrared spectra and air pollution

Fourier transform IR spectroscopy can be used to detect and measure pollutants in the air such as carbon monoxide, sulphur dioxide and ozone. It can also be used to measure the concentration of carbon dioxide in the air. The method uses the fingerprint region of the IR spectrum to identify particular molecules. To measure the concentration of carbon monoxide:

- draw the polluted air through a sample chamber
- a beam of infrared radiation is continuously passed through the sample chamber and a reference chamber (with no carbon monoxide (CO) present). Any decrease in intensity of the beam at a particular wavenumber is due to the presence of carbon monoxide
- a detector measures the difference in IR radiation between the two chambers and the amount of CO recorded automatically
- since different pollutants have characteristic IR spectra, with maximum absorbance at particular wavenumbers, this method can measure several pollutants at the same time.

Key points

- Particular groups can be identified from infrared spectra by their typical wavenumbers.

- Typical wavenumbers for functional groups are 1680–1750 cm⁻¹ for aldehydes and ketones and 3580–3650 cm⁻¹ for the O—H group in alcohols.

- A wide peak in the 2800–3500 cm⁻¹ region indicates hydrogen bonding in alcohols or carboxylic acids.

- Infrared spectroscopy is used in monitoring air pollutants such as carbon monoxide and sulphur dioxide. The concentration of carbon dioxide in the air can also be measured.

Learning outcomes

On completion of this section, you should be able to:

- explain the basic principles of a mass spectrometer (including a block diagram)

- use mass spectral data to determine relative isotopic masses and abundances

- describe how mass spectra are used to distinguish between molecules of similar relative molecular mass.

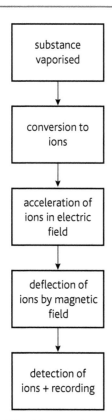

Figure 9.7.2 *Main stages in a mass spectrometer*

The mass spectrometer

The **mass spectrometer** can be used to measure relative atomic masses accurately and to identify organic compounds. Figure 9.7.1 shows a mass spectrometer and Figure 9.7.2 shows a block diagram of the main stages involved.

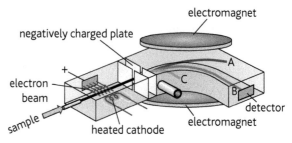

Figure 9.7.1 *A mass spectrometer*

The main stages are:

- Vaporisation of the sample.
- Ionisation: high-energy electrons from a heated cathode collide with atoms or molecules of the sample and knock out one or more of the electrons from the sample. Positive ions are formed.
- The ions are accelerated by an electric field (through a negatively-charged plate).
- The ions are deflected (bent) by a magnetic field.
- The ions are detected and recorded.

For a given electric and magnetic field, only those ions with a particular charge and mass hit the detector. By gradually increasing the strength of the magnetic field, ions of increasing **mass/charge ratio** (m/z ratio) hit the detector. (z is the charge on the ion, usually $+1$.) In Figure 9.7.1, if B represents an ion with a mass charge ratio of 15, an ion heavier than this, e.g. m/z 16, is deflected less (line A) and an ion lighter than this, e.g. m/z 14, is deflected more (line C). If the ion is doubly charged, it is deflected twice as much. A $^{208}Pb^+$ ion has an m/z ratio of 208, so a $^{208}Pb^{2+}$ ion has an m/z ratio of 104.

A mass spectrum shows the **relative abundance** (relative amount) of each ion plotted against the m/z ratio. Figure 9.7.3 shows the mass spectrum of germanium.

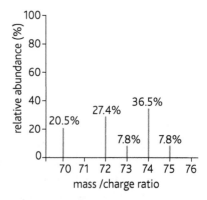

Figure 9.7.3 *Mass spectrum of germanium showing the % abundance of each peak*

Accurate atomic and molecular masses

Relative atomic masses

Mass spectra can be used to identify the different isotopes present in an element. Different isotopes are detected at particular whole-number m/z ratios because each proton and neutron has a relative mass of 1.

Mass spectra can be used to calculate relative atomic masses:

Step 1: Multiply each isotopic mass by its % abundance.

Step 2: Add the figures together.

Step 3: Divide by 100.

Worked example

Using the information in Figure 9.7.3, the relative atomic mass of germanium can be calculated as follows:

Step 1: $(20.5 \times 70) + (27.4 \times 72) + (7.8 \times 73) + (36.5 \times 74) + (7.8 \times 75)$

Step 2: $1435 + 1972.8 + 569.4 + 2701 + 585 = 7263.2$

Step 3: $7263.2 / 100 = 72.632$

Notice that the relative atomic mass is the weighted mean of the masses of all the mass numbers of the isotopes present.

Molecular masses from mass spectra

The mass spectrum of chlorine (Figure 9.7.4) shows peaks due to singly-charged ions of the ^{35}Cl and ^{37}Cl isotopes. The small peaks at 17.5 and 18.5 are due to the doubly-charged ions $^{35}Cl^{2+}$ and $^{37}Cl^{2+}$.

The spectrum will also show small peaks caused by ionised chlorine molecules, Cl_2^+. There are 3 peaks of these molecular ions:

m/z 70: due to ^{35}Cl—^{35}Cl m/z 72 due to ^{35}Cl—^{37}Cl
m/z 74 due to ^{37}Cl—^{37}Cl

These small peaks due to the molecular ions are called the **molecular ion peaks**.

Did you know?

A high-resolution mass spectrometer can help distinguish between molecules of apparently the same relative molecular mass. This is because it can measure relative isotopic masses very accurately.

For example, SO_2 and S_2 both have a relative molecular mass of approximately 64 (the relative atomic mass of sulphur is 32 and of oxygen is 16).

A high-resolution mass spectrometer can measure SO_2 and S_2 more accurately to 63.962 and 63.944, respectively (^{16}O = 15.995 and ^{32}S = 31.972).

Key points

■ Mass spectrometry involves converting gaseous atoms to ions, accelerating the ions, deflecting the ions in a magnetic field and detecting the ions.

■ A mass spectrum of a sample of an element shows the relative abundance of isotopes plotted against the mass/charge ratio.

■ Relative atomic mass is the weighted mean of the isotopic masses.

■ High-resolution mass spectrometry can distinguish between molecules with similar relative molecular mass.

✓ Exam tips

1 If you are asked to calculate relative atomic masses given relative abundances rather than % abundances, you have to add up the total abundance then divide by this number in Step 3.

2 In a mass spectrum, the peak height gives the relative abundance of each isotope.

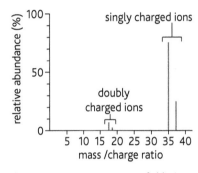

Figure 9.7.4 *Mass spectrum of chlorine (up to m/z = 40)*

On completion of this section, you should be able to:

- use data from mass spectra to distinguish between molecules of the same relative molecular mass
- explain how fragmentation patterns give information about the nature of molecules
- predict the identities of simple organic molecules
- explain the terms 'base peak' and 'molecular ion'
- explain the significance of the $M+1$ peak in mass spectra.

Mass spectra of compounds

When we ionise molecular compounds such as propanone in a mass spectrometer, a single electron may be removed from the molecule. The peak arising from this ionisation is called the **molecular ion peak**, M^+. It gives the molecular mass of the molecule. For propanone, the molecular ion peak will appear at a mass/charge ratio (m/z ratio) of 58 (see Figure 9.8.1). In many compounds, the relative abundance of the molecular ion peak is usually very low. This is because the molecule breaks up in the mass spectrometer to form fragments having particular m/z ratios. This process is called **fragmentation**.

- Fragmentation generally occurs where the bonds are weakest.
- The more stable the fragment, the greater is its abundance in the mass spectrum.
- Tertiary carbocations, e.g. $(CH_3)_3C^+$ tend to be more stable than secondary carbocations, e.g. $CH_3CH^+CH_3$ and the secondary more stable than the primary, e.g. $CH_3CH_2^+$.

Figure 9.8.1 shows the mass spectrum of propanone.

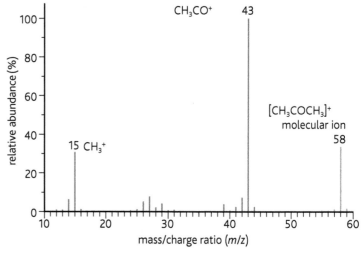

Figure 9.8.1 Mass spectrum of propanone

Each fragment has a particular m/z ratio. Notice the large peaks at 15 and 43 m/z ratios. The peak at 15 is due to a CH_3^+ ion $(C + 3H = 12 + (3 \times 1) = 15)$. The peak at 43 is due to a CH_3CO^+ ion $(2C + 3H + O = ((2 \times 12) + (3 \times 1) + 16 = 43))$. These ions have been formed in the following way:

The relative abundance of the fragments is compared with the tallest peak in the spectrum which is given an abundance of 100%. This peak is called the **base peak**. For propanone, the base peak is at m/z 43.

Distinguishing between molecules

Butane and methylpropane have the same molecular mass, 58. We can use mass spectrometry to distinguish between these molecules. The spectra are shown in Figure 9.8.2.

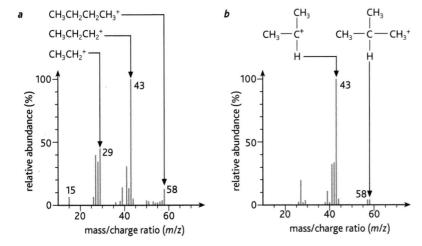

Figure 9.8.2 *Mass spectrum of **a** butane $CH_3CH_2CH_2CH_3$ and **b** methylpropane $(CH_3)_3CH$*

Notice the differences in the spectra especially at m/z 29:

In butane m/z 29 is due to $CH_3CH_2^+$

In methylpropane there is no m/z 29 peak.

So there is no $CH_3CH_2^+$ in methylpropane.

In butane m/z 43 is due to $CH_3CH_2CH_2^+$

In methylpropane m/z 43 is due to $CH_3C^+HCH_3$

In both butane and methylpropane m/z 58 is the molecular ion peak.

The $M+1$ peak

Figure 9.8.3 shows the mass spectrum of ethanol. The very small peak 1 m/z unit beyond the molecular ion peak is called the **$M+1$ peak**.

The $M+1$ peak arises because in any organic compound 1.10% of the carbon atoms are of the ^{13}C isotope. We can work out the number of carbon atoms (n) in a molecule by using this fact.

$$n = \frac{100}{1.10} \times \frac{\text{abundance of } M+1 \text{ ion}}{\text{abundance of molecular ion, } M}$$

So, if the molecular ion peak has an abundance of 49.3% and the $M+1$ peak has an abundance of 3.8%, the number of carbon atoms in the compound is:

$$n = \frac{100}{1.10} \times \frac{3.8}{49.3} = 7\text{C atoms}$$

Did you know?

If an organic compound contains Cl atoms you can get a $M+2$ peak and a $M+4$ peak in the spectrum. This is because Cl has two isotopes ^{35}Cl and ^{37}Cl. For example in the compound CH_2Cl_2, an $M+2$ peak is due to $^{35}ClCH_2{}^{37}Cl^+$ and an $M+4$ peak is due to $^{37}ClCH_2{}^{37}Cl^+$.

Did you know?

Two uses of mass spectrometry are:

- testing the urine of athletes for the presence of drugs
- monitoring pollutants in river water.

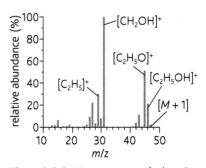

Figure 9.8.3 *Mass spectrum of ethanol*

Key points

- In a mass spectrometer, compounds break up into fragments.

- The mass spectrometer can be used to identify organic compounds from their typical fragmentation patterns.

- The molecular ion peak is produced by the loss of one electron from a molecule.

- The molecular mass of a compound can be determined from its molecular ion peak.

- Mass spectra can be used to distinguish between compounds of the same molecular mass.

- The $M+1$ peak can be used to deduce the number of carbon atoms in a compound.

Revision questions

Answers to all revision questions can be found on the accompanying CD.

$$h = 6.63 \times 10^{-34}\,\text{J s}; \quad c = 3.0 \times 10^{8}\,\text{m s}^{-1};$$
$$\text{Avogadro's number} = 6.02 \times 10^{23}$$

1 The diagram below shows some of the energy levels of a hydrogen atom.

$$n=3 \quad\underline{\hspace{4cm}}\quad -2.40 \times 10^{-19}\,\text{J}$$

$$n=2 \quad\underline{\hspace{4cm}}\quad -5.48 \times 10^{-19}\,\text{J}$$

$$n=1 \quad\underline{\hspace{4cm}}\quad -21.8 \times 10^{-19}\,\text{J}$$

a An electron moves from energy level $n = 3$ to energy level $n = 1$, emitting a photon. What is the energy change for this process?

b What is the frequency of the emitted radiation?

c What would be the total energy change, **in kJ**, for one mole of hydrogen atoms (each with one electron), where the same electronic transition occurs?

2 a By comparing the energy of an X-ray photon ($\lambda_x = 6 \times 10^{-11}\,\text{m}$) with that of an infrared photon ($\lambda_{IR} = 5 \times 10^{-6}\,\text{m}$), explain why short length radiation is more damaging to human tissue than longer wavelengths.

b What is the energy of a photon of red light ($\lambda = 680\,\text{nm}$)?

c A sodium lamp radiates 15 W of yellow light ($\lambda = 590\,\text{nm}$).

  i What is the energy of each photon emitted?

  ii How many photons are emitted from the lamp per second? ($1\,\text{W} = 1\,\text{J s}^{-1}$)

d Determine the frequency of:

  i an X-ray beam which has a wavelength of $4.88\,\text{Å}$

  ii an ultraviolet ray with wavelength of 211 nm

  iii microwaves with a wavelength of 0.211 cm ($1\,\text{Å} = 10^{-10}\,\text{m}; 1\,\text{nm} = 10^{-9}\,\text{m}$)

e Given that the difference in energy between the 3p and 3s orbitals is $3.38 \times 10^{-19}\,\text{J}$, what is the wavelength of radiation (in m and nm), that would be absorbed if an electron moves from the 3s to a 3p orbital?

3 Determine the absorbance of the following solutions:

a a solution with a transmittance of 0.314

b a solution with a % transmittance of 42.4

c a solution with molar absorptivity of 10 000 and concentration of $3.25 \times 10^{-5}\,\text{mol dm}^{-3}$, where the absorbance is measured in a 1.0 cm cell.

4 a i If a solution of an analyte in water with a concentration of $1.00 \times 10^{-4}\,\text{mol dm}^{-3}$ is examined at λ_{max} 220 nm, the absorbance is found to be 1.40. If the path length of the cell is 1.0 cm, what is the molar absorptivity of the analyte at this wavelength?

  ii This analyte has another absorption band at λ_{max} 268 nm. If the same solution is examined at 268 nm, ($\varepsilon = 900$), what will be the absorbance reading?

b A student is planning to record the UV spectrum of an analyte which has $\lambda_{max} = 310\,\text{nm}$ ($\varepsilon = 24\,000$). What concentration should be prepared in order to obtain an absorbance of 0.512 at the maximum, if a cell with a 1.00 cm path length is used?

c A solution of an analyte with a concentration of $2.10 \times 10^{-3}\,\text{mol dm}^{-3}$ has an absorbance of 0.455, when measured in a cell with a 1.00 cm path length. This solution is then diluted and the absorbance measured in the same manner, is found to be 0.184. What is the concentration of the diluted solution?

5 A UV-visible absorption spectrometer measures the absorbance of specific substances, and from these measurements, the concentration can be determined by applying Beer–Lambert's law.

a Identify one factor that can cause a deviation from Beer–Lambert's law.

b Explain why the use of complexing agents is sometimes required in this form of spectroscopy.

c The absorption spectra is actually due to the presence of chromophores. What is meant by the term 'chromophore'?

d State two limitations of UV-visible spectroscopy.

e A solution of $KMnO_4$ has an absorbance value of 0.508, when measured at a wavelength of 525 nm. What is the transmittance of this solution? What is the % transmittance of this solution?

f The molar absorptivity of $KMnO_4$ is 2240 at this wavelength. If the absorbance of the solution is measured in a cell with a 1.50 cm path length, what is the concentration in **mol dm⁻³**, **g dm⁻³** and **ppm**? (Molar mass $KMnO_4$ is 158 g mol⁻¹; concentration in ppm = mass of solute (mg) / volume of solution (dm³)).

6 Explain each of the following in terms of electronic transitions:

  a The molecule $CH_3CH_2CH_2CH_3$ does not absorb in the UV-visible region above 200 nm, whereas the molecule $CH_2{=}CHCH{=}CH_2$ does.

  b The molecule

  has two absorption peaks in the accessible UV-visible region: $\lambda_{max} = 320$ ($\varepsilon = 21$) and $\lambda_{max} = 213$ ($\varepsilon = 7080$).

  c λ_{max} in the UV region is higher for the molecule $CH_2{=}CHCH{=}CHCH{=}CH_2$ than for the molecule $CH_2{=}CHCH{=}CHCH_2CH_3$.

7 a Explain why a molecule such a CH_3I would absorb radiation in the IR region, but the molecule I_2 would not.

  b Explain which of the bonds C–Cl or C–I would show a stronger absorption.

  c Briefly explain how a sample is prepared for IR spectroscopy.

  d The IR spectrum is a plot of % transmittance against wavenumber.

   i Define the term 'wavenumber'.

   ii Calculate the wavenumber corresponding to a
   ■ frequency of 2.7×10^{13} Hz
   ■ wavelength of 8.18×10^{-6} m.

8 a How could IR spectroscopy be used to distinguish between the isomers in **i** and in **ii**?

   i CH_3CH_2OH and CH_3OCH_3

   ii cyclohexane and hex-2-ene

  b Predict the positions of the major absorption bands in the IR spectra of the following:

   i Cyclohexanol

   ii

  c The IR spectrum of the compound

  has major absorption peaks in the regions 3350–3500, 3000–2850 and 1600–1459. Suggest the identity of the groups responsible for these absorptions. (Refer to Section 9.6.)

9 a Explain how each of the following processes occurs in the mass spectrometer:

   i vaporisation

   ii ionisation

   iii acceleration

   iv deflection

   v detection.

  b Identify **two** ways in which data from a mass spectrum can be used.

  c Explain the meaning of these terms which are used in the analysis of a mass spectrum:

   i the m/z ratio

   ii the $M + 1$ peak

   iii the base peak.

10 a Determine the relative atomic mass for each element given in **i–iii**.

   i The relative atomic mass of Fe, where the isotopes ^{54}Fe, ^{56}Fe, ^{57}Fe and ^{58}Fe have **relative abundances** of 5.85 %, 91.75 %, 2.12% and 0.280% respectively.

   ii The relative atomic mass of Li where the isotopes ^{6}Li and ^{7}Li have the **ratio of peak intensities** of 0.082 : 1.00 respectively.

   iii The relative atomic mass of Mg where the isotopic abundances of ^{24}Mg, ^{25}Mg and ^{26}Mg are 0.79, 0.10 and 0.11 respectively.

  b The mass spectrum of a straight chain alkane is obtained where the molecular ion peak has a relative abundance of 12% and the $M+1$ peak has a relative abundance of 0.55%.

   i How many carbon atoms are there in a molecule of this compound?

   ii What is the molecular formula for this compound?

   iii Write the structural formula for this compound.

   iv Predict four major fragments, along with their m/z values, that would appear in the mass spectrum of this compound.

  c Some of the peaks of the mass spectrum for the compound butan-2-ol have m/z values of 74, 59, 45, 29 and 15. Suggest the identity of the species responsible for these peaks.

11 An atom emits radiation of wavelength 1.5×10^{-7} m. Calculate the energy of this radiation per mole of atoms. $c = 3.0 \times 10^8 \, m\,s^{-1}$.
Planck constant = 6.63×10^{-34} J s. Avogadro number, 6.02×10^{23}.

10.1 Introduction to chromatography

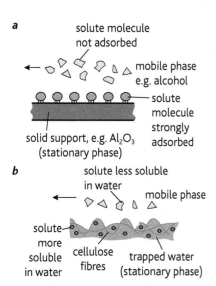

Figure 10.1.1 a *Adsorption chromatography;* **b** *Partition chromatography*

The theory of chromatography

Introduction

Chromatography is a technique used to separate the **components** of a mixture. In chemistry, a component is one of the compounds or elements in a mixture. Chromatography works by dividing the components of a mixture between two different phases. We call this **partitioning**. Water and hexane do not mix. They form two separate layers. When we shake an aqueous solution of iodine with hexane, most of the iodine goes into the hexane layer but some remains in the aqueous layer. We say that the iodine has been partitioned between the two layers (see Section 10.8).

Did you know?

The word *chromatography* was taken from two Greek words meaning 'colour writing'. In the early days of chromatography the technique was only used to separate coloured substances.

Adsorption and partition chromatography

There are two main types of chromatography, adsorption chromatography and partition chromatography. They both depend on partitioning the components of a mixture between two phases, a stationary phase and a mobile phase.

The **stationary phase** can be a solid, e.g silica (silicon(IV) oxide) or alumina (aluminium oxide) or a liquid, e.g. water trapped between cellulose fibres. The stationary phase tends to hold back the components of the mixture which are attracted to it. The greater the attraction, the slower is the movement of the components during chromatography.

The **mobile phase** can be a liquid or a gas. The greater the solubility of a particular component in the mobile phase, the faster is the movement of that component during chromatography.

During chromatography, the mobile phase moves over the stationary phase. The different components in the mixture are attracted to the stationary phase to different extents. So some move faster than others in the mobile phase and separation occurs.

Adsorption chromatography: The stationary phase is a solid. **Adsorption** is the process of forming bonds of varying strength with a solid surface. As the mobile phase moves over the solid (stationary phase) some components are adsorbed more strongly to the solid than others and separation occurs.

Partition chromatography: The stationary phase is a liquid surrounding particles of a solid support. As the mobile phase moves over the stationary phase, the components which are more soluble in the liquid stationary phase will be held back and move more slowly than those that are more soluble in the mobile phase. This difference depends on the partition coefficient (see Section 10.8).

Four types of chromatography

Column chromatography

This is a form of adsorption chromatography. Silica, alumina or a resin (stationary phase) is mixed with a solvent such as alcohol or water (mobile phase). The mixture is packed into a column. A mineral wool or sintered glass plug at the bottom of the column keeps the stationary phase in place. The mixture to be separated is added to the top of the tube and then allowed to soak into the stationary phase. Solvent (mobile phase) is then continuously added to the top of the tube. The solvent moves through the tube separating the components of the mixture. See Section 10.2 for more on column chromatography procedure.

Thin-layer chromatography (TLC)

This is a form of adsorption chromatography. Silica, alumina or cellulose (stationary phase) are made into a paste and spread in a thin even layer over a glass or plastic plate. A small spot of the mixture to be separated is put near the bottom of the plate. The plate is placed in a solvent (mobile phase) and the solvent allowed to move up the stationary phase. As the solvent moves up the stationary phase, the spot separates into its components. See Section 10.2 for more on **thin-layer chromatography** procedure.

Paper chromatography

This is a form of partition chromatography. The stationary phase is water absorbed onto the cellulose of the paper. The mobile phase is usually an organic liquid or a mixture of solvents. The apparatus (see Section 10.2) is similar to that for TLC but with paper instead of a thin layer of stationary phase. As the solvent moves up the paper, the components partition themselves between the water (stationary phase) and the solvent (mobile phase). See Section 10.2 for more on paper chromatography procedure.

Gas–liquid chromatography (GLC)

This is a form of partition chromatography. The stationary phase is a high-boiling point liquid, e.g. a long-chain hydrocarbon oil supported on a porous inert solid such as silica or alumina. The mobile phase is an unreactive gas, e.g. nitrogen, helium, argon. This is called the carrier gas. The mixture to be separated is injected into the apparatus (see Figure 10.2.5, page 111). As the mixture is carried through the apparatus by the gas, those substances that are more soluble in the oil will travel more slowly and those that are less soluble will travel faster. See Section 10.2 for more on **gas–liquid chromatography** procedure.

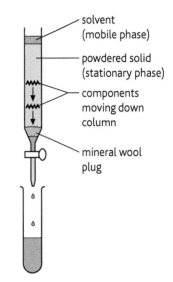

Figure 10.1.2 A chromatography column

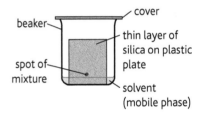

Figure 10.1.3 Thin-layer chromatography

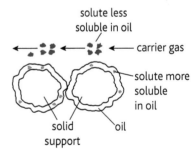

Figure 10.1.4 Partition in gas–liquid chromatography

Key points

- In chromatography the mobile phase moves solutes through or over a stationary phase.

- Chromatographic separation occurs when a solute is transferred from the mobile phase to the stationary phase by either partition between a gas and a liquid or adsorption on a solid surface.

- The stationary phase can be a solid, e.g. silica or alumina or a liquid.

- The mobile phase can be a liquid or a gas.

- Paper, thin-layer, column and gas–liquid chromatography are used to separate substances.

Learning outcomes

On completion of this section, you should be able to:

- understand the terms 'retention factor' and 'retention time', 'visualising agent' and 'solvent front'

- describe the basic steps involved in separating the components of a mixture using chromatography

- understand how R_f values and retention times are used in quantifying substances.

Column chromatography

After the column has been packed, the procedure is:

- Add the mixture to be separated to the top of the column.

- Allow the mixture to soak into the column (open the tap at the bottom).

- Keep adding solvent (mobile phase) to the top of the column and let the solvent drain through carrying the components of the mixture with it. Don't let the column run dry.

- Collect fractions of appropriate volume in test tubes at the bottom of the column.

These may be from $1\,cm^3$ to $100\,cm^3$ depending on the size of the column.

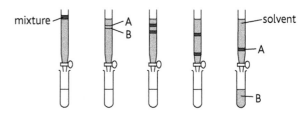

Figure 10.2.1 *Collecting the fractions in column chromatography. The diagram shows the separation of a mixture containing two components, A and B.*

Thin-layer and paper chromatography

The procedure is the same for both TLC and paper chromatography.

- Place a spot of the mixture to be separated on the base line.

- Dip the paper (or TLC plate) into a solvent. The solvent level must be below the base line.

- Allow the solvent to move up the paper (or TLC plate).

- When the **solvent front** is near the top, mark its position.

✅ Exam tips

When carrying out TLC or paper chromatography remember:

1 Mark the base line and solvent front in *pencil*. The components of ink will be chromatographed as well!

2 Put a cover over the tank or beaker. This prevents solvent loss by evaporation and allows equilibration of the vapour with the liquid solvent.

Visualising agents and retention value, R_f

When the components of a mixture are not coloured, we spray the finished chromatogram or TLC plate with a **visualising agent** (locating agent). This reacts with the colourless components. A coloured compound is formed. The colour may sometime need to be 'developed' by warming the treated paper. Different types of visualising agents are used for different types of compound, e.g. ninhydrin reacts with amino acids to give purple coloured spots.

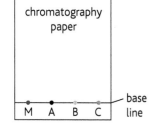

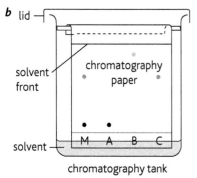

Figure 10.2.2 *Paper chromatography; **a** A mixture and three pure substances, A, B, C are placed on the base line; **b** The chromatogram after separation. The mixture contains A and C.*

We can identify the components on a chromatogram by comparing how far the spots have moved from the base line compared with how far the solvent front has moved. We call this the **retention value**, R_f.

$$R_f = \frac{\text{distance from base line to centre of spot}}{\text{distance from base line to solvent front}}$$

In Figure 10.2.4 the R_f value of component A is 4/6 = 0.67 and the R_f value of component B is 1.5/6 = 0.25.

Gas–liquid chromatography, GLC

The apparatus for GLC is shown in Fig 10.2.5. The procedure is:

- The mixture to be separated is injected into the gas which flows through a long spiral tube containing the stationary phase. The substance injected must be able to form a vapour easily, e.g. it has to be a gas, liquid or volatile solid.
- The time of injection is recorded.
- The components of the mixture separate in the tube.
- The components leaving the tube are detected (usually by measuring changes in thermal conductivity of the gas coming out from the apparatus).

The separated components leave the tube at different times. The time between injection and detection is called the **retention time**. We can identify a component by matching its retention time with known retention times for particular substances under the same conditions.

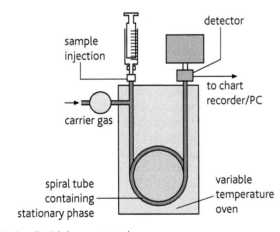

Figure 10.2.5 Gas–liquid chromatography

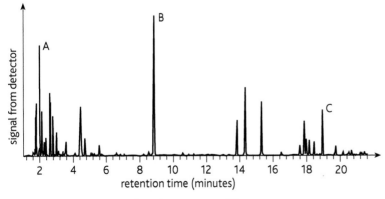

Figure 10.2.6 A GLC trace. Each peak represents a different component.

Did you know?

Paper chromatography can also be carried out by placing the solvent in a trough at the top of the apparatus. This as called descending paper chromatography.

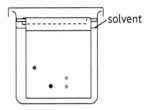

Figure 10.2.3

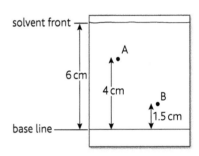

Figure 10.2.4 Calculating R_f values from a TLC plate

Key points

- In paper and thin-layer chromatography, the components of the mixture are identified by their R_f values.
- The R_f value is the distance moved by a specific component from the base line divided by the distance moved by the solvent front from the base line.
- A visualising reagent reacts with colourless components in a paper or thin-layer chromatogram to give the separated components colour.
- In gas–liquid chromatography, the components are identified by retention times.

Did you know?

Columns for column chromatography can be as wide as 3.2 metres and as high as 15 metres. These can be used for separating and purifying large amounts of materials.

Column chromatography

This method has the advantage that fairly large amounts of material can be separated, e.g. mixtures of amino acids or mixtures of proteins. Columns can be made of various sizes. Larger-sized columns can be used for purifying natural products, e.g. plant oils such as limonene or for purifying drugs. Each fraction collected from the bottom of the column can be analysed automatically, e.g. proteins can be analysed quantitatively by UV-spectrometry by measuring the absorbance at 280 nm. Amino acids can be analysed quantitatively by reaction with ninhydrin and measurement of the colour intensity using visible-spectroscopy. Figure 10.3.1 shows a typical analysis of separate amino acids collected from the column.

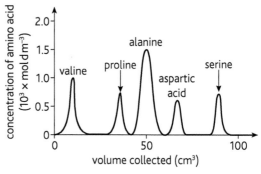

Figure 10.3.1 *Quantifying amino acids separated during column chromatography. The amino acid content in each tube was determined by colorimetry.*

Thin-layer and paper chromatography

These methods can be used to separate small amounts of compounds such as amino acids, plant pigments and food colourings. We cannot, however, completely separate compounds with similar R_f values. We can use two-dimensional chromatography to help overcome this problem. After the initial chromatography, we allow the paper to dry and then turn it 90° and carry out chromatography in a different direction using a different solvent (Figure 10.3.2).

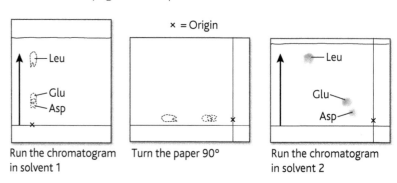

Figure 10.3.2 *Two-dimensional paper chromatography of the three amino acids leucine (Leu), glutamic acid (Glu) and aspartic acid (Asp)*

The R_f values of coloured materials are easily calculated. If colourless compounds are separated, the paper or TLC plate must be sprayed with a suitable visualising agent. These methods are useful when small amounts of materials are to be identified but are less useful when quantification of large amounts of materials are required. Quantification can, however, be made by cutting out the spots, removing the compound in the spot with a solvent, then quantitatively analysing the solution using UV-visual spectroscopy or mass spectrometry.

Gas–liquid chromatography, GLC

The components arising from GLC can be fed directly into a mass spectrometer or IR spectrometer for identification. The method can be very sensitive and is used to separate and identify traces of substances in foodstuffs and analyse pesticide concentrations in the environment. It is often used in forensic analysis to separate and identify particular compounds, in medicine to determine gas concentration in blood samples and in analysing fuels. A well-known use is in testing the blood or urine of athletes for performance-enhancing drugs. The retention times of the components are matched with those of known substances using the same flow rate, carrier gas and stationary phase. Additional confirmation is given by mass spectroscopy. One limitation of GLC is that similar compounds have similar retention times.

Quantifying amounts using GLC

A typical gas chromatogram is shown in Figure 10.3.3.

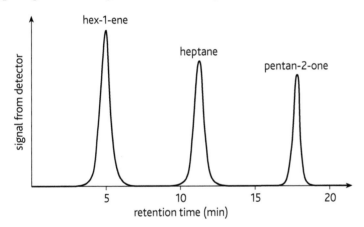

Figure 10.3.3 The separation of three volatile organic compounds by gas–liquid chromatography

If spectroscopic methods are not available for quantifying the amount of each component, we can use the area under each peak as a rough guide. Since the peaks are roughly triangular in shape, we can easily calculate the area:

area of triangle $= \frac{1}{2} \times$ base $\times$ height

The computer attached to the GLC apparatus can calculate this value quite easily. We can also calculate the percentage (%) composition of any component in the mixture as long as:

- each peak is well separated
- there are peaks for all the components in the mixture
- the peak area is directly proportional to the concentration.

$$\text{\% composition of component A} = \frac{\text{peak area of A}}{\text{sum of the peak areas of all the components}}$$

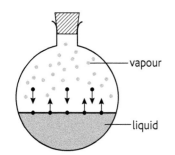

Figure 10.4.1 Liquid–vapour equilibrium in a flask of water

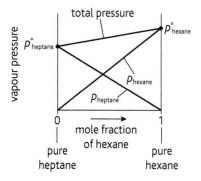

Figure 10.4.2 Vapour pressure–composition curve for a mixture of hexane and heptane

Did you know?

All liquid mixtures deviate to a greater or lesser extent from Raoult's law. Those that more or less follow Raoult's law are called zeotropic mixtures. The word *zeotropic* comes from two Greek words meaning 'to boil' and 'state'.

Vapour pressure

Figure 10.4.1 shows a flask of water from which air has been removed. Water molecules evaporate from the surface and form a vapour. Eventually, an equilibrium is established, in which the rate of movement of water molecules to and from the liquid is the same.

$$H_2O(l) \rightleftharpoons H_2O(g)$$

The pressure exerted by the vapour molecules is called the **vapour pressure**. Vapour pressure varies with temperature. In a mixture, the pressure exerted by each component in the vapour alone is called the **partial vapour pressure**.

Mixtures of liquids – Raoult's law

- Liquids that dissolve in each other, whatever the volumes added, are said to be **miscible**. An example is a mixture of ethanol and water.

- Liquids that do not dissolve in each other are said to be **immiscible**. Immiscible liquids form separate layers after shaking them together. An example is a mixture of hexane and water.

We say the mixture is an **ideal solution** if it obeys **Raoult's law**:

The partial vapour pressure of a component in a mixture equals its mole fraction, x_A, × the vapour pressure of the pure component.

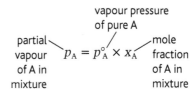

For an ideal mixture of two components A and B, the total vapour pressure is:

$$p_T = p_A + p_B = (p^o_A \times x_A) + (p^o_B \times x_B)$$

Mixtures form ideal solutions if:

- the intermolecular forces between the molecules in the mixture are similar to those in the pure substances

- there are no enthalpy or volume changes on mixing.

Figure 10.4.2 shows the effect of Raoult's law on an ideal solution (a mixture of hexane and heptane) at a fixed temperature.

As the mole fraction of the more volatile component (hexane) increases:

- the partial pressure of the more volatile component increases and the partial pressure of the less volatile component decreases

- the total pressure increases so that at any point the sum of the partial pressures = total pressure

- the line for the total pressure is a straight line.

The experimental measurement of vapour pressure is difficult. So we often use boiling points of mixtures to mirror vapour pressure–composition graphs. The boiling point is the temperature at which the vapour pressure equals the atmospheric pressure. The molecules of a more volatile component of a mixture escape more readily into the vapour phase. So the more volatile component has a lower boiling point. For an ideal mixture, the boiling points vary as shown in Figure 10.4.3. There is a linear relationship with the mole fraction of each component.

Non-ideal mixtures

Some mixtures do not obey Raoult's law. We say they deviate from Raoult's law. The deviations can be positive or negative. There is not a linear relationship between the vapour pressure or boiling point and composition.

✓ Exam tips

You will *not* be expected to do calculations based on vapour pressure–composition curves or boiling point–composition curves.

Positive deviations from Raoult's law

Example: ethanol and cyclohexane.

- The bonding between ethanol and cyclohexane is weaker than between ethanol alone and cyclohexane alone.
- The cyclohexane molecules get in the way of the hydrogen bonding in the ethanol molecules. There is net bond breaking.
- The molecules in the mixture are more likely to escape from the liquid to the vapour.
- So the vapour pressure of the mixture is higher than expected for an ideal mixture.
- The boiling point is therefore lower than expected.

Negative deviations from Raoult's law

Example: ethyl ethanoate and trichloromethane.

- The bonding between ethyl ethanoate and trichloromethane is stronger than between ethyl ethanoate and trichloromethane alone.
- There is net bond forming.
- The molecules in the mixture are less likely to escape from the liquid to the vapour.
- So the vapour pressure of the mixture is lower than expected for ideal mixture.
- The boiling point is therefore higher than expected.

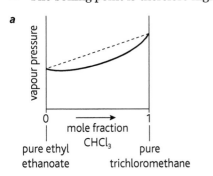
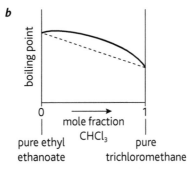

Figure 10.4.5 *Negative deviation from Raoult's law. The dashes show the line expected if Raoult's law is obeyed.* **a** *Vapour pressure–composition curve and* **b** *boiling point–composition curve.*

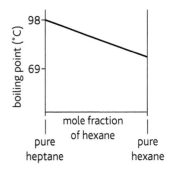

Figure 10.4.3 *Boiling point–composition curve for a mixture of hexane and heptane*

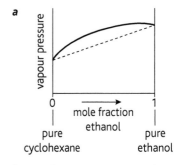

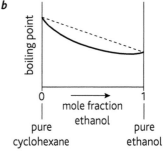

Figure 10.4.4 *Positive deviation from Raoult's law. The dashes show the line expected if Raoult's law is obeyed.* **a** *Vapour pressure–composition curve and* **b** *boiling point–composition curve.*

Key points

- Raoult's law states that the partial vapour pressure of a component in a mixture equals the vapour pressure of the pure component × its mole fraction.
- An ideal mixture obeys Raoult's law.
- Boiling point–composition curves for ideal mixtures can be drawn which show a linear relationship between the composition and the boiling point.
- Boiling point–composition curves for non-ideal mixtures may show positive or negative deviations.

On completion of this section, you should be able to:

- understand the principles on which distillation and fractional distillation are based.

Simple distillation

When we need to separate a product which is a liquid or solid with a boiling point below about 250 °C from a mixture, we can use simple distillation. For compounds boiling above about 180 °C we use an air condenser.

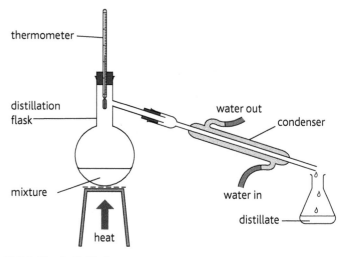

Figure 10.5.1 *Simple distillation*

The mixture is boiled. The vapour of the component of the mixture with a lower boiling point vaporises then liquefies in the condenser and is collected in the receiver. If the other components of the mixture have sufficiently higher boiling points, their vapours will condense in cooler parts of the flask and not reach the condenser.

This method can be used if the boiling points of the components of the mixture are different enough so they do not to reach the condenser at the same time as each other. An example is the purification of sea water from the minerals present in it. The water distils off first, leaving the mineral salts in a more concentrated solution.

Fractional distillation

Simple distillation cannot be used if the distillate is found to contain other compounds with similar boiling points to the one required. In order to separate liquids within a narrow range of boiling points we use **fractional distillation**.

A column containing glass beads or rods or a column with bulbous surfaces is used (Figure 10.5.2). The column allows the ascending vapour to come into contact with the descending liquid and separation occurs through successive liquid vapour equilibria.

The longer the column and the slower the heating, the better is the separation of the liquids whose boiling points are close to one another.

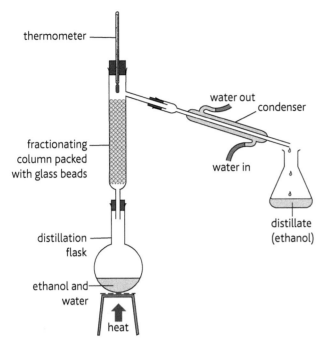

Figure 10.5.2 *Fractional distillation*

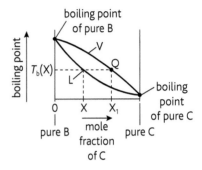

Figure 10.5.3 *Boiling point composition curve for a mixture of liquids B and C. The line L represents the boiling point of the liquid. The line V represents the boiling point of the vapour.*

How does fractional distillation work?

To see how distillation works we can refer to Figure 10.5.3.

The two components of the mixture are B and C. At its boiling point (when the vapour pressure = external pressure), the mole fraction of the C in the mixture is X. The vapour, however, has more of the volatile component, C, in it. The vapour at this temperature is shown by Q. The composition of this distillate is shown by X_1. In other words, the mole fraction of C in the vapour has increased.

Figure 10.5.4 shows what happens as we continue to distil the mixture.

When we heat up the mixture of composition X to its boiling point, T_b:

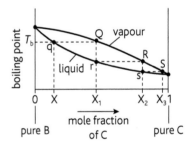

Figure 10.5.4 *Boiling point composition curve for fractional distillation*

- The vapour gets richer in the more volatile component, C.
- At this boiling point, T_{b1}, the vapour and liquid are in equilibrium. This is shown by the line qQ.
- The vapour rises up the column and gets cooler until it condenses.
- This temperature drop is represented by line Qr.
- At r, where the boiling point is lower, there is a new equilibrium with the mole fraction of C being X_1.
- At this lower boiling point, T_{b2}, the vapour and liquid are in a new equilibrium. This is shown by the line rR.
- The composition of this new mixture is X_2, which has an even greater mole fraction of component C.
- The process continues in this way until the vapour consists of C alone (mole fraction = 1).

As the vapours rise in the column through successive equilibria, they become richer and richer in the more volatile component. Eventually they pass out of the column and are condensed. The liquid in the flask gradually gets richer and richer in the less volatile component. Its boiling point increases.

Key points

- Simple distillation is used to separate substances with boiling points which are far apart from each other, e.g. water from mineral salts.

- Fractional distillation is used to obtain a complete separation of liquids within a narrow range of boiling points.

- Fractional distillation works because a series of equilibria occur further up the column, in which the vapour gets increasingly rich in the more volatile component.

10.6 Azeotropic mixtures and other distillations

Learning outcomes

On completion of this section, you should be able to:

- interpret boiling point–composition curves of azeotropic mixtures in a qualitative manner

- compare the efficiency of the separation of ethanol from water by simple and fractional distillation

- understand the advantage of carrying out distillation under reduced pressure.

Azeotropic mixtures

Some compositions of liquid mixtures which deviate widely from ideal behaviour can be separated or partly separated by fractional distillation. However, some other compositions of liquid mixtures which deviate widely from ideal behaviour cannot be separated by fractional distillation. These mixtures are called **azeotropic mixtures**. Examples are HCl and water, ethanol and water, and ethanol and benzene. They cannot be separated because the boiling point composition diagram shows a distinct maximum or minimum. Figure 10.6.1 shows the boiling point composition curve for a mixture of ethanol and benzene. Ethanol is the more volatile component. A mixture having a minimum boiling point is formed.

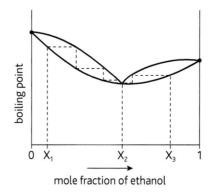

Figure 10.6.1 *Boiling point–composition curve for a mixture of benzene and ethanol*

If we start with a mixture of initial mole fraction of ethanol X_1 and apply the same ideas about fractional distillation as in Section 10.5, fractional distillation results in:

- pure benzene and
- a mixture having composition X_2.

The mole fraction of ethanol represented by X_2 is 0.54.

If we start with a mixture of initial mole fraction of ethanol X_3, fractional distillation results in:

- pure ethanol and
- a mixture having composition X_2.

The mole fraction of ethanol represented by X_2 is again 0.54. The minimum boiling point is 67.8 °C.

The mixture of benzene and ethanol represented by X_2 is called an azeotropic mixture or a **constant boiling mixture**. You can see that we cannot completely separate such mixtures by fractional distillation. Azeotropic mixtures can also have maximum boiling points. An example is a mixture of trichloromethane and propanone (Figure 10.6.2).

Ethanol distillation

When we distil an ethanol–water mixture using simple distillation, we do not get pure ethanol in the receiver. We get a mixture of water and ethanol. The process is not very efficient. We would have to repeat this many times to get a very high alcohol content. This does not matter too

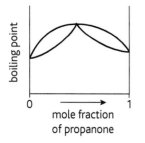

Figure 10.6.2 *Boiling point–composition curve for a mixture of trichloromethane and propanone*

much if we are making alcoholic beverages such as rum or whisky because the alcohol content does not need to be 100%. Ethanol can be made more concentrated by fractional distillation, which is a more efficient process. But even fractional distillation does not produce pure alcohol because alcohol forms a constant boiling mixture with water. Fractional distillation of a mixture of ethanol and water produces a constant boiling mixture with an ethanol content of 95.6 mole %. In order to produce pure ethanol the remaining water is removed by shaking the alcohol–water mixture with silica gel. This absorbs the water but not the ethanol.

Distillation under reduced pressure

This method is used to distil liquids which would either completely or partly decompose if they are distilled at atmospheric pressure (1 atm). It is sometimes used in preference to steam distillation (see Section 10.7) because it can sometimes produce a purer product. The method is also called **vacuum distillation**. If the apparatus is attached to a water pump, the pressure can be reduced to the value of the vapour pressure of water (about 0.016 atm). The apparatus used is shown in Figure 10.6.3.

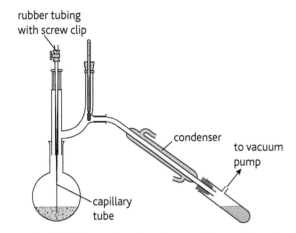

10.6.3 *Apparatus for distillation under reduced pressure. The capillary tube reduces sudden movements of the liquid ('bumping') which often occurs in these distillations.*

When the pressure is reduced, the boiling point of the liquid is also reduced. We can purify phenylamine (boiling point 184 °C at 1 atm) by distilling impure phenylamine under reduced pressure. The boiling point of phenylamine is reduced to 77 °C if distillation is carried out at about 0.02 atm pressure.

The solvent dimethyl sulfoxide, $(CH_3)_2SO$, is commonly purified by vacuum distillation because it is more stable when distilled under reduced pressure.

Did you know?

The phrase '70% proof' applied to alcoholic drinks does not mean that the drink contains 70% alcohol. It comes from an ancient test so that authorities could collect tax on 'proof spirit'. The drink was poured over gunpowder. If the gunpowder could be lit, the spirit was 'proof'.

Key points

- An azeotropic mixture cannot be completely separated by distillation.
- Azeotropic mixtures form constant boiling mixtures which have either a minimum or maximum boiling point depending on the mixture.
- Fractional distillation is more efficient than repeated simple distillations.
- Distillation under reduced pressure is used to distil compounds which may decompose if distilled at atmospheric pressure.

Steam distillation

Immiscible liquids do not mix. An example is plant oils and water. For two immiscible liquids the total vapour pressure is equal to the sum of the vapour pressures of the pure components:

$$p_T = p^o_A + p^o_B \qquad \text{e.g.} \qquad p_T = p^o_{H_2O} + p^o_{oil}$$

So the total vapour pressure is higher than the vapour pressure of either component alone. The temperature at which the mixture boils is therefore lower than that of either component alone. We can take advantage of this to distil plant oils or other substances which are immiscible with water and have other contaminating compounds in them. This process is called **steam distillation**. When steam is bubbled through a liquid which is immiscible with water, the vapour pressure is increased and the mixture boils at a temperature lower than 100 °C (Figure 10.7.1).

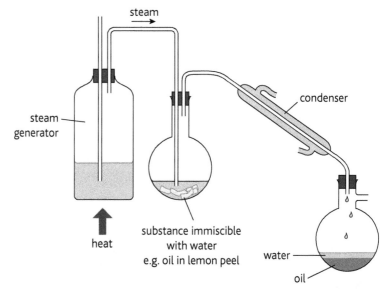

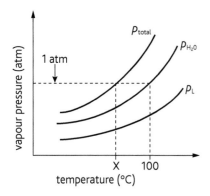

Figure 10.7.1 *Vapour pressure–boiling point curves for two immiscible liquids, L and water. The boiling point of the mixture, X, is below that of water.*

10.7.2 *Steam distillation of a plant oil*

We can apply steam distillation to distil a plant oil such as limonene from lemon or orange peel. The procedure is:

- Cut the lemon peel into small sections.
- Put the peel into a flask and add just enough water to cover the peel.
- Bubble steam through the mixture.
- Condense the mixture of steam and plant oil.
- Collect the plant oil/water mixture in a receiver.
- Use a separating funnel (see Section 10.8) to separate the plant oil from the water.

Steam distillation is used to purify compounds such as phenylamine, $C_6H_5NH_2$ and nitrobenzene, $C_6H_5NO_2$, which have relatively high boiling points. Distillation at a lower temperature reduces the risk of thermal decomposition.

Calculations involving steam distillation

We can calculate the molar mass of a liquid from steam distillation by comparing the mass of water and liquid present in the distillate. Since each component of the mixture in steam distillation behaves independently, we can use Raoult's law and the equation $p_T = p^o{}_{H_2O} + p^o{}_A$ to develop a third equation:

$$\frac{p_{H_2O}}{p_A} = \frac{n_{H_2O}}{n_A}$$

p are partial pressures of liquid A and water

n are the number of moles of A and water.

Since $n = \dfrac{\text{mass of liquid } (m)}{\text{molar mass of liquid } (M)}$

we can write this equation as:

$$\frac{p_{H_2O}}{p_A} = \frac{m_{H_2O}/M_{H_2O}}{m_A/M_A}$$

Worked example 1

When nitrobenzene is steam distilled at a pressure of 1 atm (760 mm mercury), the mixture distils at 99 °C. At this temperature, the vapour pressure of water is 733 mm mercury. The distillate contains 40 g of water and 10 g of nitrobenzene. Calculate the relative molecular mass of nitrobenzene. ($M_{water} = 18$).

Step 1: Calculate the vapour pressure of nitrobenzene:
760 – 733 = 27 mm mercury.

Step 2: Substitute the figures into the equation:

$$\frac{p_{water}}{p_{nitrobenzene}} = \frac{m_{water}/M_{water}}{m_{nitrobenzene}/M_{nitrobenzene}}$$

$$\frac{733}{27} = \frac{40/18}{10/M_{nitrobenzene}}$$

So $M_{nitrobenzene} = \dfrac{733}{27} \times \dfrac{18 \times 10}{40} = 122$

Worked example 2

Phenylamine was steam-distilled at 98.6 °C and a pressure of 760 mm mercury (1 atm). The vapour pressure of water at this temperature was 720 mm mercury. The distillate contained 25 g water. Calculate the mass of phenylamine in the distillate.

$$(M_{water} = 18, M_{phenylamine} = 93)$$

Step 1: Calculate the vapour pressure of phenylamine:
760 – 720 = 40 mm mercury.

Step 2: Substitute the figures into the equation:

$$\frac{p_{water}}{p_{phenylamine}} = \frac{m_{water}/M_{water}}{M_{phenylamine}/M_{phenylamine}}$$

$$\frac{720}{40} = \frac{25/18}{m_{phenylamine}/93}$$

So $m_{phenylamine} = \dfrac{25}{18} \times \dfrac{40}{720} \times 93 = 7.2\,g$

Key points

- Steam distillation is used to separate mixtures of immiscible liquids or to extract oils from plants.

- Steam distillation is used to purify high boiling point liquids such as phenylamine and nitrobenzene.

- For two immiscible liquids the total vapour pressure = sum of the vapour pressures of the pure components.

- Relative molecular masses can be calculated from the moles, n, of each immiscible liquid in the distillate by the equation:

$$\frac{p_{water}}{p_A} = \frac{n_{water}}{n_A}$$

10.8 Solvent extraction

Introduction

Solvent extraction can be used to separate two solutes dissolved in a solvent. A second solvent which is immiscible with the first is used to extract one of the solutes from the first solvent. The extraction is carried out using a separating funnel (see Figure 10.8.1). For example, if we have an aqueous solution of iodine (very slightly soluble in water) and sodium chloride (soluble in water) and we want to separate the iodine we:

- Put the aqueous mixture in the separating funnel then add another solvent which is
  - immiscible with water
  - a good solvent for iodine, e.g. hexane.
- Shake the contents of the separating funnel then let the layers settle. The iodine will go into the component in which it is more soluble and the sodium chloride remains in the aqueous layer.
- Run off the bottom layer.
- Repeat the process with fresh solvent until nearly all the iodine has been removed from the aqueous layer.
- The solvent (hexane) is evaporated, leaving the iodine as a solid.

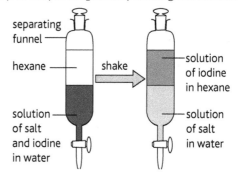

Figure 10.8.1 *When an aqueous solution of salt and iodine is shaken with hexane, the iodine moves to the hexane layer*

The distribution coefficient

Solvent extraction depends on the relative solubility of a solute in two immiscible solvents. The amount of solute which is partitioned (divided) between two immiscible solvents depends on the equilibrium concentrations present. For example, if we shake aqueous ammonia with an equal volume of trichloromethane, $CHCl_3$, in a separating funnel, ammonia molecules move between the two layers until an equilibrium is established.

$$NH_3(CHCl_3) \rightleftharpoons NH_3(aq)$$

The concentration of ammonia in each layer can be determined by titration. The equilibrium constant for this process is called the **distribution coefficient** or **partition coefficient**, K_D. The value of K_D may vary with temperature. Most values are quoted at 298 K.

For this equilibrium $K_D = \dfrac{[NH_3(aq)]}{[NH_3(CHCl_3)]} = 23.3$

Note that partition coefficients are usually quoted for the equilibrium expression which has the higher concentration on the upper line.

Calculations using the distribution coefficient

Worked example 1

A solution of butanedioic acid (BDA) in ether contains 4.0 g BDA in 20 cm³ of ether. This solution is shaken with 50 cm³ of water.

Calculate the mass of BDA extracted into the water layer.

$$K_D \text{ for BDA(ether)} \rightleftharpoons \text{BDA(water) is 6.7}$$

Step 1: Calculate the concentration in each component.

$$[\text{BDA(water)}] = \frac{m}{50} \times 1000 \qquad [\text{BDA(ether)}] = \frac{(4.0 - m)}{20} \times 1000$$

(if m is the mass in the water, this must have come from the ether layer so m is subtracted from the mass in the ether layer)

Step 2: Substitute values into the equilibrium expression.

$$K_D = \frac{[\text{BDA(water)}]}{[\text{BDA(ether)}]} \qquad 6.7 = \frac{\dfrac{m}{50}}{\dfrac{(4.0 - m)}{20}}$$

$$m = 3.8\,g$$

Worked example 2

Several extractions using small portions of solvent are more efficient than using a single larger volume of solvent. We can see this by comparing two solvents A and B, where K_D for [B]/[A] = 4.

■ If we have 15 g of solute in 50 cm³ of A and shake it with 50 cm³ of B, the amount of solute transferred to B is given by:

$$4 = \frac{m}{50} \div \frac{(15 - m)}{50} = 12\,g$$

■ If we extract the solute using two portions of 25 cm³ the first extraction gives:

$$4 = \frac{m}{50} \div \frac{(15 - m)}{25} = 10\,g$$

and the second extraction gives:

$$4 = \frac{m}{50} \div \frac{(5 - m)}{25} = 3.3\,g$$

So the total extracted is 10 + 3.3 = 13.3 g

Some uses of solvent extraction

■ Ether extraction is often used in chemistry to separate products of organic synthesis from water. The ether layer floats above the aqueous layer. The ether layer is separated and the ether (very flammable) is left to evaporate leaving the organic product behind. The technique is especially useful when the product required is volatile or unstable to heat.

■ Cosmetic manufacturers often have to take account of the relative solubility in the different solvents they use in the formulation of cosmetic creams and hair colourings.

■ In metallurgy ideas of solvent extraction are used to determine how impurities are distributed between molten metals and solid metals.

■ Environmental chemistry uses ideas of relative solubility when deducing how readily pollutants are taken up into the groundwater from industrial wastes.

■ The food industry uses solvent extraction to analyse the relative solubility of compounds in fats and aqueous solutions.

☑ *Exam tips*

In Step 2, you can ignore the 1000 relating to the concentration as long as the units of volume are the same. This is because the 1000 cancels. So, as long as the units of volume are the same, the expression can be simplified like this.

☑ *Exam tips*

Remember that if you are calculating the mass of solute extracted using two extractions, do not use the same amount of solute for the second extraction as you used in the first. A lot of solute will have been removed in the first extraction. You should use the amount of solute in the first extraction MINUS the amount removed in the first extraction.

Key points

■ Solvent extraction is based on the relative solubility of a solute in two immiscible liquids.

■ The distribution coefficient is the constant at a particular temperature which relates the concentration of a solute partitioned between two immiscible solvents.

■ The amount of solute extracted from a particular solution by another solvent can be calculated if the distribution coefficient is known.

■ Several extractions using small portions of solvent are more efficient than using a single larger volume of solvent.

10.9 Distillation and solvent extraction: applications

Learning outcomes

On completion of this section, you should be able to:

- give examples of the application of distillation in the petroleum industry, the rum industry and the fragrance industry

- understand how acids and bases can be separated by solvent extraction

- select appropriate methods of separation when given the physical and chemical properties of the components of a mixture.

Did you know?

One of the earliest references to rum (1651) is in a document from Barbados. 'The chief fuddling (drink) they make in the island is Rumbullion ... this is made of sugar canes distilled, a hot, hellish and terrible liquor'.

Distillation in the rum industry

Rum is made from molasses (a by-product of sugar refining) or from sugar cane juice.

Yeast and water are added to the molasses (which contains the carbohydrates for the yeast to feed on) to start the fermentation. Different producers use particular strains of yeast but many use the foam from previous fermentations.

$$C_6H_{12}O_6(aq) \rightarrow 2C_2H_5OH(aq) + 2CO_2(g)$$

When fermentation is complete the aqueous mixture is distilled.

Some producers work in batches (pot stills). Heat is applied directly to the pot still and the alcohol and aroma compounds giving the rum its characteristic flavours evaporate.

Other producers use column distillation. This is a type of fractional distillation. The column behaves like a series of pot stills. It is filled with materials that give a large surface area for condensation of the rising vapours. The rising vapour which is higher in alcohol than the fermentation mixture condenses in the higher levels of the column, which are cooler. The vapours become progressively richer in alcohol the higher up the still they reach (see Section 10.5).

The fragrance industry

Many fragrances are hydrocarbon-based compounds often with aromatic rings. Many are oily in nature. Some plants contain only very small amounts of volatile oils or the oils are easily denatured if temperature is too high. Either solvent extraction or steam distillation is commonly used to extract fragrances.

In **solvent extraction** the raw material is immersed in the solvent and shaken for a time to extract the aromatic compound. Hexane and dimethyl ether are two of the solvents most commonly used. If the plant contains a lot of water, two layers are formed – an aqueous layer (arising from the water present in the plant material) and an organic layer which contains the desired mixture of compounds. Some fragrances can be extracted from the waxy 'concrete' which remains after solvent extraction or distillation. These can be extracted with ethanol, e.g. jasmine fragrance and rose oil. Ethanol is not usually used to extract fresh plant material as it is soluble in water.

Steam distillation is used for extracting oils and fragrances from flowers, leaves and stems of plants, since the oil distils off with the water at temperatures lower than 100°C. The temperature of distillation is sufficiently low to prevent decomposition of the fragrance molecules.

Other methods

Vacuum distillation: This can be used to extract some components which are easily denatured if fractional distillation is used.

Fractional distillation: If the fragrance molecules are stable, this is used to remove contaminating molecules with less pleasant odours.

Petroleum distillation

Crude oil is a mixture of hydrocarbons. It is separated into a number of different components by several types of distillation:

- Some of the dissolved gases are removed by simple distillation.
- Primary distillation then separates the oil into several groups of components (light distillate, middle distillate and residue) by fractional distillation.
- Petroleum fractionation is discussed further in Section 12.1.

Using solvents to extract acids and bases

Ionic salts are generally soluble in water. Neutral molecules, especially larger organic molecules tend to be insoluble in water.

In a mixture of an organic acid such as phenol and an organic base such as phenylamine, the following equilibria exist:

$$C_6H_5OH \rightleftharpoons C_6H_5O^- + H^+ \quad \text{and} \quad C_6H_5NH_2 + H^+ \rightleftharpoons C_6H_5NH_3^+$$

If we add a slightly stronger base to a mixture of phenol and phenylamine, the phenylamine remains uncharged but the phenol becomes deprotonated. The phenol is present in the form of an ionic salt but the phenylamine is present only as molecules.

$$C_6H_5OH + OH^- \rightarrow C_6H_5O^- + H_2O$$
$$C_6H_5NH_3^+ + OH^- \rightarrow C_6H_5NH_2 + H_2O$$

If we add an acid, the phenol remains in its molecular form but the phenylamine becomes protonated. The phenylamine is in the ionic form.

$$C_6H_5O^- + H^+ \rightarrow C_6H_5OH$$
$$C_6H_5NH_2 + H^+ \rightarrow C_6H_5NH_3^+$$

We can separate the ionic form from the molecular form by solvent extraction:

- A mixture of organic acid and organic base is dissolved in a suitable solvent e.g. dichloromethane or ethoxyethane (an ether).
- The mixture is poured into a separating funnel.
- An aqueous solution of another acid (or base) is added to adjust the pH so that the molecular or ionic form you want is present.
- The water/solvent mixture is shaken and the phase containing the compound in either the molecular form (in non-aqueous solvent) or ionic form (in the aqueous layer) is separated off.

Selecting a suitable method of separation

This depends on:

- *Solubility:* An insoluble solid can be separated from a soluble solid by dissolving the latter then filtering.
- *Boiling points:* A mixture of liquids with different boiling points can be separated by fractional distillation.
- *Solubility in different solvents:* Use solvent extraction.

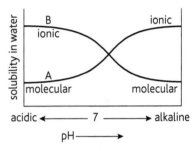

Figure 10.9.1 *The molecular and ionic forms of organic acids (A) and bases (B) at different pH values*

Key points

- Distillation by batch process or fractional distillation is used in the production of spirits such as rum and whisky.
- Many perfumes and fragrances are extracted using steam distillation or solvent extraction.
- Acids and bases can be extracted or purified using solvent extraction.
- The selection of a suitable method of separating substances depends on properties such as boiling points and solubility in particular solvents.

Answers to all exam-style questions can be found on the accompanying CD

$h = 6.63 \times 10^{-34} \, J\,s; \, c = 3.0 \times 10^8 \, m\,s^{-1}$

Multiple-choice questions

1 The diagram shows some of the energy levels of a hydrogen atom. An electron moves from energy level $n = 3$ to energy level $n = 1$, emitting a photon. What is the frequency of the emitted radiation?

$n = 3$ ——————————— -2.40×10^{-19} J

$n = 2$ ——————————— -5.48×10^{-19} J

$n = 1$ ——————————— -21.8×10^{-19} J

A 8.27×10^{14} Hz   B 2.93×10^{15} Hz

C 3.29×10^{15} Hz   D 3.65×10^{15} Hz

2 What is the energy of a light photon with a wavelength of 650 nm?

A 3.40×10^{-36} J   B 1.44×10^{-29} J

C 3.06×10^{-19} J   D 6.63×10^{-19} J

3 There are four types of electromagnetic radiation in the list below:

i   X-ray   ii   infrared

iii   microwaves   iv   ultraviolet

Which of the lists below gives the radiation in order of increasing wavelengths?

A i, iv, ii, iii   B ii, iv, iii, i

C iv, iii, i, ii   D iii, i, ii iv

4 For a system where a solute **R** is distributed between two solvents, and **R** has the same molecular form in both solvents, the equilibrium process may be represented as:

$$R_{\text{solvent 1}} \rightleftharpoons R_{\text{solvent 2}}$$

Which of these factors would affect the value of the partition coefficient?

i   the mass of the solute originally in solvent 1

ii   the volume of the solvents

iii   the temperature

A i and ii   B ii and iii

C i, ii and iii   D only iii

5 The partition coefficient for Br_2 between water and tribromomethane at 25 °C is 66.7, where the bromine is in the same molecular form in both solvents and has a higher solubility in tribromomethane. Which of these statements is true?

i   If the concentration of Br_2 in $CHBr_3$ is 0.250 mol dm^{-3}, then the concentration in H_2O is 3.75×10^{-3} mol dm^{-3}.

ii   The equilibrium system can be represented by the equation

$$Br_{2(H_2O)} \rightleftharpoons Br_{2(CHBr_3)}$$

iii   At a temperature above 25 °C, the value of the partition coefficient would change from 66.7.

A only i   B i and ii

C ii and iii   D i, ii and iii

6 The actual amount of potassium ions present in a food sample was 4.25%. The food sample was analysed by four different experimenters, using different approaches, who repeated each test four times. The results are given below. Which experimenter's results show the highest degree of accuracy and precision?

A 5.21, 5.22, 5.21, 5.21, 5.23

B 5.45, 4.11, 4.25, 4.84, 6.43

C 4.25, 4.24, 4.23, 4.26, 4.27

D 4.20, 4.15, 4.35, 4.30, 4.25

7 In a back-titration, 25.0 cm^3 of 1.00 mol dm^{-3} HCl was added to an antacid containing the active agent $CaCO_3$ in a conical flask. The excess HCl was titrated with 0.55 mol dm^{-3} NaOH solution and it was found that a volume of 27.3 cm^3 was required for a complete reaction. How many moles of the active agent $CaCO_3$, does the antacid contain?

A 1.5×10^{-2} B 5.0×10^{-3}

C 2.5×10^{-2} D 7.5×10^{-3}

8 A 1.5234 g sample of impure $BaCO_3$ was reacted with excess dilute hydrochloric acid. The CO_2 liberated was absorbed by NaOH and found to weigh 0.1425 g. What is the % of barium in the sample? [Relative atomic mass C = 12.01; O = 16.00; Ba = 137.33]

A 41.94 %   B 9.35 %

C 29.19 %   D 90.65 %

9 A solution of $KMnO_4$ of molarity 0.102 mol dm^{-3} required 30.35 cm^3 to react completely with 22.24 cm^3 of solution X of molarity 0.348 mol dm^{-3}. What is the mole ratio for the reaction between MnO_4^- and X?

A 1 MnO_4^- : 2 X   B 2 MnO_4^- : 5 X

C 0.4 MnO_4^- : 1 X   D 5 MnO_4^- : 2 X

10 Propanone shows absorption maxima at λ_{max} 189 nm and λ_{max} 279 nm. What type of transition is responsible for each of these absorptions?

λ_{max} 189 nm	λ_{max} 279 nm
A $\pi \rightarrow \pi^*$	$n \rightarrow \pi^*$
B $\sigma \rightarrow \pi^*$	$\sigma \rightarrow \sigma^*$
C $n \rightarrow \pi^*$	$n \rightarrow \sigma^*$
D $n \rightarrow \pi^*$	$\sigma \rightarrow \pi^*$

Structured questions

11 a Distinguish between the terms 'gravimetric analysis' and 'titrimetric analysis'. [2]

b A sample was prepared by mixing a number of substances and the entire sample was then dissolved in distilled water in a 250 ml volumetric flask. Two students, Julia and Jenny, were then asked to determine the mass of calcium present in the 5.524 g sample by using different approaches. Julia titrated three 25.0 ml aliquots of the solution with 0.100 mol dm^{-3} EDTA and found that the mean titre was 37.10 cm^3. Jenny placed 150 ml in a beaker and added excess Na_2CO_3 solution. She then collected, dried and weighed the precipitate, which had a mass of 2.403 g.

[Relative atomic mass Ca = 40.078; O = 15.999; C = 12.011]

i Based on Julia's approach, what is the mass of Ca in the sample? [4]

ii Calculate the % of calcium in the sample from Julia's approach. [1]

iii What was the identity of the precipitate obtained by Jenny? [1]

iv Based on Jenny's approach, what is the mass of Ca in the sample? [4]

v The actual mass of Ca in the sample prepared was 1.518 g. From a consideration of the relative error, comment on the accuracy of Julia's and Jenny's results. [2]

vi Suggest ONE factor that could have caused the mass obtained by Jenny to be higher than the expected value. [1]

12 a Explain what is meant by the *vapour pressure* of a liquid. [2]

b i How does an increase in external pressure affect the boiling point of a liquid? [1]

ii Explain why this occurs. [2]

c For the two-component system acetone and carbon disulphide, an azeotrope is formed at mole fraction 0.36 of acetone, and has the boiling point of 312 K. The boiling point of pure acetone is 329 K and the boiling point of pure carbon disulphide is 319 K.

i Sketch the boiling point composition curve for this mixture. [3]

ii Explain what happens to the:
- intermolecular forces of attraction
- total volume of the liquid

when these two liquids are mixed. [2]

iii Would the mixing be an endothermic or exothermic process? [1]

iv What would be the identities of the first few drops of distillate and the liquid remaining in the distillation flask, if the following mixtures are fractionally distilled?
- 0.79 mole fraction of acetone [2]
- 0.25 mole fraction acetone. [2]

13 a UV-visible spectroscopy and IR spectroscopy are two of the spectroscopic methods of analysis. Compare these methods in terms of the type of information that can be obtained from their use. [4]

b The concentration of $FeSCN^{2+}$ in a solution is to be determined by UV-visible spectroscopy, at λ_{max} 580 nm and using a 1.00 cm cell. A calibration curve was plotted for the system and the molar absorptivity coefficient was found to be 7.00×10^3 dm^3 mol^{-1} cm^{-1}. Five students each diluted the original solution, by placing 10.0 cm^3 in a 100 cm^3 volumetric flask and adding distilled water up to the graduation mark. The absorbance of the each diluted solution was then determined and the results are shown in the table below.

Student	1	2	3	4	5
Absorbance at λ_{max} 580 nm	0.552	0.564	0.550	0.554	0.540

i Explain how the value of molar absorptivity coefficient, ε, was determined from the calibration curve. [1]

ii Calculate the mean absorbance and use this to determine the average concentration in mol dm^{-3} in the diluted solution. [2]

iii What is the concentration of $FeSCN^{2+}$ in the original solution? [3]

iv Find the standard deviation of the students' results. [2]

c The IR spectrum for the compound ethyl ethanoate

$$CH_3-\overset{\overset{O}{\|}}{C}-C-CH_2CH_3$$

shows absorption bands at 3000–2850 cm^{-1}, 1742 cm^{-1} and 1241 cm^{-1}.

Identify the type of bonds that most likely give rise to these bands. [3]

11 Aluminium

11.1 Locating chemical industry

Did you know?

The traditional chemical industry is often found near ports which allow easy import of raw materials and export of products. The siting of plants which produce speciality chemicals such as pharmaceutical is less affected by the factors mentioned here.

Chemical industry in the Caribbean

Cuba	Trinidad and Tobago
Nickel	Ammonia and
Petroleum	fertilisers
	Pitch
	Petroleum
	Methanol
	Iron
Jamaica	**Guyana**
Bauxite and alumina	Bauxite

Factors that influence the location of chemical industry

In recent years, there has been much discussion in the Caribbean about the building of aluminium smelting plants. There are many factors that influence the location of chemical industry, some are discussed here:

Nearness to the source of raw materials (including water supply):

- Metal ores are heavy, so to transport them a long distance is expensive and takes a lot of energy.
- Water is used as a raw material in many processes, e.g. sulphuric acid manufacture and production of ethanol. So water supplies should be close. This is why some chemical plants are located near the sea. Bromine is manufactured from seawater, so it makes sense to set up a chemical plant manufacturing bromine near the sea. In addition to being a raw material, water is used in most chemical industries for cooling, as a solvent or as a heating medium. So most chemical plants are near the sea or near rivers.

Good communications network: To transport the raw materials to the factory and the products away from the factory, good road and rail links are needed. For environmental reasons, heavy materials such as metal ores are best transported by rail rather than road, to minimise pollution. Many metal ores are transported by ship, so the nearness to a port or specially built harbour is very important.

Labour: Employees in the chemical industry tend to be more skilled than those in many other areas of manufacturing. The location of the chemical plant should be near enough to a centre of population that can provide such people.

Availability of cheap energy: Many chemical reactions can *only* be carried out at high temperatures, e.g. smelting metals. Many others need heat to give an economic rate of conversion of reactants to products, e.g. synthesis of ammonia and making sulphuric acid. Purification and separation processes also require heating, e.g. distillation.

Gas and oil are used often used as heat sources in industry. Although gas and oil can be transported, it makes more economic sense to site a factory or plant near an oil refinery or close to a gas pipeline. Most oil refineries are near the coast because the petroleum is transported from the producing countries in ships.

Iron and steel plants are often sited near deposits of coal which are used to make the coke necessary for the reduction of the iron ore.

Electrolysis of aluminium oxide requires large amounts of energy to keep the electrolyte molten. This is provided by electrical heating. Electrical energy is expensive, so aluminium smelting plants are often sited near sources of hydroelectric power or close to the coast.

Environmental and social factors:

- The plant should preferably not be built in an area of natural beauty. If it is (i) the process carried out should be in the national interest

(ii) there should be no alternative sites (iii) there should be is a plan to restore the area when the factory is no longer viable.

- It should not be sited so close to a centre of population that people are affected by the fumes or noise (i) from the factory (ii) from lorries transporting raw material and products.
- It should be sited to minimise the possibility that waste products from the factory damage the environment, e.g. waste products do not get into rivers and harm wildlife.

Political and historical factors: If there is industry already existing in the area, it makes it easier to set up a new factory because the impact on people and the environment is minimised. In the past, many factories were established near to coalfields, where there were already centres of industrial production. Political factors such as the availability of government grants and tax incentives to develop particular regions of a country also play a part in the location of a factory.

An aluminium smelter for Trinidad and Tobago?

The Government of Trinidad and Tobago wanted to set up an aluminium smelter in the south of Trinidad near the coastal town of La Brea. This would be of advantage to the economy of the country. The advantages of this location are:

- The raw materials can be sourced nearby; purified aluminium oxide can be transported from Jamaica and bauxite ore is available from Guyana.
- It is near the sea so water is readily available and raw materials can be transported by ship.
- There is already industry present in the area – the Trinidad Lake of Pitch.
- Finance was available from the Chinese company involved in the deal and a loan from the Chinese Government was also available.
- The project would provide work for many people especially as a new port and new power station to provide the huge amount of electricity required were going to be built.

This project has currently stalled however, for various reasons:

- Environmentalists are worried about the damage to the coastline and surrounding land.
- Some local people are worried about the pollution from the fumes and that the plant is too close to the town of La Brea.
- The material that lines the smelter pots needs replacing from time to time and people are worried about the dangers from the cyanide, arsenic and other toxic substances in the pots.
- The roads leading to the site are not in a suitable condition.

Large scale chemical industry in the Caribbean

Although the Caribbean has some reserves of metals such as nickel and copper, these are not present in worthwhile amounts to merit extraction on a large scale. The main large-scale chemical industry is processing bauxite.

Safety in the chemical industry

Care has to be taken to prevent:

- the leakage of harmful or poisonous reactants/products
- the leakage of vapours/waste gases which may be flammable, explosive or radioactive
- the corrosion of pipes and vessels in the chemical plant
- the weakening of the structure of pressure vessels.

Problems can be overcome by:

- access to proper safety clothing including protective suits, face masks and access to oxygen and chemicals to neutralise spillages
- regular checking for corrosion.
- checking the conditions of pressure vessels
- monitoring the working environment using specific sensors
- checking that instrumentation is working correctly.

Key points

- The location of chemical industry depends on the source of raw materials, available energy, ease of transport, and environmental and social factors.

- Bauxite and alumina production are important industries in Jamaica, ammonia/fertiliser production and petroleum/ natural gas production are important in Trinidad and Tobago.

- The chemical industry should have tight safety requirements to eliminate possibilities of fires, explosions or release of toxic materials.

Introduction

Aluminium is the most abundant metal in the Earth's crust. It forms about 7.5% of the crust and is present combined with silica in clay, shale, granite and slate. It is extracted from bauxite ores, which contain hydrated aluminium oxide, e.g. gibbsite $Al_2O_3 \cdot 3H_2O$, boehmite $Al_2O_3 \cdot H_2O$.

Aluminium is obtained from bauxite in two stages:

- Purification of bauxite to make aluminium oxide (alumina).
- Electrolysis of the purified alumina.

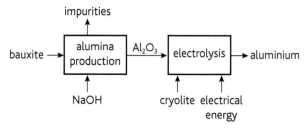

Figure 11.2.1 *Steps in the extraction of aluminium from bauxite*

The main impurities in bauxite are oxides of silicon, iron and titanium. 'White bauxite' has 1–4% silica and very little iron. 'Red bauxite' has 3–25% iron(III) oxide and 1–7% silica.

Alumina production

Sodium hydroxide is used to dissolve the aluminium oxide from the ore and separate the impurities. Approximately 2.5 tonnes of bauxite are needed to make 1 tonne of purified alumina, Al_2O_3. In order to remove the iron oxide (Fe_2O_3), titanium dioxide (TiO_2) and most of the silicon dioxide (SiO_2) present, the bauxite is treated with concentrated sodium hydroxide. Aluminium oxide is amphoteric, so it dissolves in sodium hydroxide. The basic oxides do not dissolve in sodium hydroxide and are filtered off.

Stage 1: Powdered bauxite is mixed with 10% NaOH and heated under pressure (4 atm) at 140 °C. This takes about 1–2 hours to complete.

$$Al_2O_3 + 2NaOH \rightarrow \underset{\text{sodium aluminate}}{2NaAlO_2} + H_2O$$

Silicon dioxide is removed either during or prior to this step by reaction with sodium hydroxide.

The sodium silicate formed is then removed by precipitation.

Stage 2: The sodium aluminate is soluble in sodium hydroxide but contaminating metal oxides are not. The impurities are allowed to settle and are filtered off.

Stage 3: The sodium aluminate is 'seeded' with pure aluminium oxide and agitated with air.

Slow cooling produces a precipitate of pure aluminium oxide trihydrate.

$$2NaAlO_2 + 4H_2O \rightarrow Al_2O_3 \cdot 3H_2O + 2NaOH$$

Stage 4: After 36 hours the alumina is removed by vacuum filtration then dehydrated in a rotary kiln at 1000 °C. The sodium hydroxide is recycled.

Electrolysis of pure aluminium oxide

Compounds of metals and non-metals only conduct electricity when molten or in solution. Aluminium oxide alone cannot be used as an electrolyte. Aluminium oxide has a very high melting point (2040 °C). It would be nearly impossible to maintain this temperature for a long time as it would require too much energy and special containers. Aluminium oxide is not soluble in water. However it will dissolve in a mixture of molten **cryolite**, Na_3AlF_6 and aluminium fluoride or calcium fluoride. This solution melts at about 1000 °C and so reduces the energy costs considerably. The added calcium fluoride or aluminium fluoride causes the melting point of the mixture to be lowered further than adding cryolite alone. This helps reduced energy consumption. So the electrolyte used is cryolite containing about 5% aluminium oxide and a little calcium or aluminium fluoride.

Electrolysis takes place in tanks (cells or 'pots') lined with carbon. This carbon is made by baking anthracite and pitch. The carbon lining of the tanks is the cathode (negative electrode). The anodes (positive electrodes) are blocks of carbon dipping into the molten electrolyte. Large cells have 20 anodes each about 400 mm wide. The anodes can be lowered further into the cell as the electrolysis proceeds.

The cells are linked in series. A voltage of 5 V is used and a huge current of 40 000 amps is required. About one-third of the electricity is used in electrolytic reactions and the rest is used to keep the electrolyte molten. It takes a lot of energy to keep the temperature of electrolyte at nearly 1000 °C for a long period of time. The production of 1 kg of aluminium requires about 15 kilowatt hours of electricity. For this reason, aluminium smelters are usually built where large amounts of electricity are available relatively cheaply, e.g. hydroelectric power. Many smelters have their own power station. The reactions at each electrode are uncertain. It is possible that some of Al_2O_3 ionises:

$$Al_2O_3 \rightarrow Al^{3+} + AlO_3^{3-}$$

At the cathode aluminium ions accept electrons and are converted to aluminium atoms:

$$Al^{3+} + 3e^- \rightarrow Al$$

The molten aluminium which is 99.9% pure collects at the bottom of each cell and is siphoned off.

At the anode a possible reaction is:

$$4AlO_3^{3-} \rightarrow 2Al_2O_3 + 3O_2 + 12e^-$$

This is often simplified as a reaction in which oxide ions form oxygen by loss of electrons: $2O^{2-} \rightarrow O_2 + 4e^-$

During the electrolysis, the oxygen produced reacts with the carbon anode which gets 'burnt away' and therefore has to be replaced periodically. The carbon dioxide formed is led away through fume hoods.

The electrolysis process is continuous. Regular additions of alumina and calcium or aluminium fluoride is required to maintain constant electrolyte composition. These additions are necessary because when the concentration of aluminium oxide falls, dangerous fluorine gas is evolved and the voltage required for the electrolysis rises.

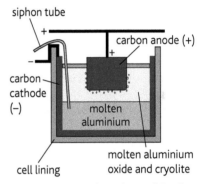

Figure 11.2.2 *An electrolysis cell for the production of aluminium*

Key points

- Bauxite is purified by dissolving the ore in sodium hydroxide, precipitating the impurities and then precipitating the purified alumina from the solution.

- Aluminium is extracted from a mixture of molten aluminium oxide in cryolite using electrolysis cells with carbon anodes and cathodes.

- Cryolite is added to dissolve the alumina. This lowers the melting point of the mixture.

- Large amounts of energy are needed to keep the cryolite–alumina electrolyte molten.

Uses of aluminium

The uses of aluminium reflect its physical and chemical properties:

Low density: Aluminium has lower density compared with most other metals. Its mass is one-third of that of steel of the same volume. So it is used where there is energy-saving advantages, e.g. (as an alloy) for aircraft fuselages, in car bodies and in parts of ships. Aluminium ladders are much lighter than wooden ones. High-tension electricity cables between pylons are usually made of aluminium, since it is less dense and cheaper than copper. These cables have a steel core because aluminium would break under its own weight if used alone.

Good strength/ mass ratio: The strength of pure aluminium to stretching (tensile strength) is 7–11 megapascals. Alloying the aluminium can increase its tensile strength 10 to 50 times. So aluminium is useful for making aircraft, ladders and lightweight cars and lorries.

Malleability and ductility: Aluminium is easily shaped, so it can be used for drinks cans and roofing materials. Aluminium foil is more flexible than foils of many other metals of similar price so can be used for food packaging where it has the additional advantages that it is non-toxic and relatively cheap. Aluminium is readily machined, cast and drawn into wires.

Good electrical conductivity: Although not as good a conductor of electricity as copper, gold and silver (it has 59% the conductivity of copper), aluminium is comparatively cheap. Together with its low density, its good electrical conductivity makes it useful for electrical wiring, especially in overhead power lines.

Good thermal conductivity: Aluminium is used to make parts of boilers, cookers and cookware because of its good thermal conductivity.

High reflectivity: Aluminium reflects light very well (92% of the light is reflected). It is one of the few metals that has excellent reflectance when it is finely powdered. This is why it is used in silver-coloured paints, mirror reflectors and heat-resistant clothing for fire fighting.

Non-magnetic: Aluminium is used in navigational equipment because it is not corroded and is not affected significantly by magnetic fields.

Corrosion resistance: Aluminium is resistant to corrosion because a thin surface layer of aluminium oxide forms on the freshly made metal when it is exposed to air. Aluminium oxide is not very reactive. So the oxide layer prevents corrosion. Unlike iron, the oxide layer does not flake off. Some alloys of aluminium are less corrosion resistant if the alloying metal is less reactive than aluminium. Because of its resistance to corrosion, aluminium is used to make window frames, roofing materials and car and plane bodies. It is also useful for drinks cans and food containers whose contents are acidic.

Good reducing agent: Aluminium is used in the chemical and steel industries as a reducing agent.

Aluminium industry and the environment

Aluminium is a very useful metal and the production of aluminium from bauxite employs many thousands of people throughout the world. Hundreds of thousands of others are employed in industries which use aluminium for making such different things as cars, electrical cables, drinks cans and window frames. But producing aluminium has an environmental cost:

Quarrying the bauxite ore

- Quarrying may destroy land which can either be used for agriculture or is an area of natural beauty, e.g. forest or hilly area.
- The quarries are unsightly, noisy and can produce dust and fumes from the explosives and vehicles used to extract and transport the ore.
- Waste rocks from the ore may produce unsightly spoil heaps.

Production of alumina

- Breaking up the bauxite ore produces dust and fumes.
- The waste products precipitated from the reaction with sodium hydroxide produces a residue which is saturated with sodium hydroxide. This 'red mud' can drain into the soil and then get into waterways. Plants and animals are poisoned by the high concentrations of sodium hydroxide and other materials present in the sludge.
- The kilns (furnaces) used to remove the water from the hydrated alumina produce the greenhouse gas, carbon dioxide and a certain amount of dust is also produced.

Electrolysis of alumina

- The electrolysis of alumina uses vast amounts of electricity. Considerable quantities of the greenhouse gas carbon dioxide are therefore formed in producing the electricity.
- The reaction of the carbon anodes with the oxygen also produces the greenhouse gas, carbon dioxide.
- The 'pots' (electrolysis cells) used to produce aluminium have a finite lifetime. When broken up or recycling the walls/ electrodes, dust is produced which may contain cyanides, arsenic and other toxic compounds.
- Fluorine gas is produced more and more as the amount of aluminium oxide in the cells decreases. Fluorine is a toxic gas which is very reactive.
- Perfluorocarbons are produced during the electrolysis due to reaction between fluorine and the carbon electrodes. Perfluorocarbons are very powerful greenhouse gases.

Did you know?

The price of metals and metal ores on the world market can have a very bad effect on individual countries and on individuals. In the early 1980s there was a world recession. The price of metals and ores dropped rapidly. This affected Jamaica badly because 75% of its exports were connected with bauxite. The falling prices led to 25% of the workforce connected with this business losing their jobs.

Key points

- The properties of aluminium make it useful for making overhead electricity cables, window frames, roofing, car and aeroplane bodies, mirrors, cookware and drinks cans.
- The properties of aluminium that are especially useful are its low density, its corrosion resistance, high strength mass ratio and high reflectance.
- Quarrying, the production of alumina and the electrolysis of alumina have particular affect on the environment by producing dust noise and materials which are toxic (or potentially toxic).
- The production of aluminium results in the production of carbon dioxide and other greenhouse gases.

Learning outcomes

On completion of this section, you should be able to:

- describe the method used in separating the components of crude oil

- understand how fractional distillation separates crude oil into its component fractions

- state the uses of the fractions obtained from crude oil as fuels and as raw materials for the petrochemical industry.

Crude oil and its components

Crude oil (petroleum) is a mixture of hydrocarbons. It contains alkanes, cycloalkanes and aromatic compounds whose relative molecular masses range from 16 to more than 400. Some of the dissolved natural gases are first removed by simple distillation. This 'stripping' is often done near the oil wells. The crude oil is then transported to an oil refinery where it undergoes fractional distillation. This separates the oil into different **fractions**. Each fraction has a particular range of boiling points e.g. the kerosene fraction has components with boiling points between 160–250 °C, while the light gas-oil fraction has components with boiling points between 250–300 °C.

The names of the different fractions and their uses are shown in Figure 12.1.1, which also shows where they come off from the distillation column.

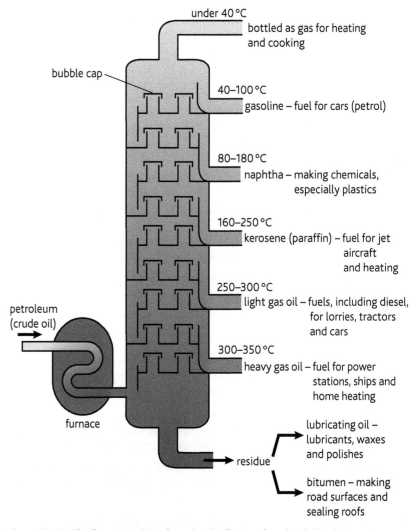

Figure 12.1.1 *The fractions arising from the distillation of crude oil. The diagram is simplified and does not show all the reflux pipes in the tower.*

The table below shows the approximate number of carbon atoms and boiling point range of some of the fractions.

Fraction	gas	gasoline	naphtha	kerosene	gas oil	residue
Boiling points /°C	below 40	40–100	80–180	160–250	250–350	above 350
Number of C atoms	1–4	4–8	5–12	10–16	16–25	above 25

The naphtha fraction is especially important. Further treatment of this leads to the formation of unsaturated compounds which are important in chemical syntheses, e.g. ethene for making poly(ethene).

Separating the fractions

The crude oil is heated in a furnace at about 400 °C. The vapour is then fed into a fractionating tower (column) which contains about 40 'trays' containing bubble caps (see Figure 12.1.1). These allow thorough mixing of the vapour with descending liquid. In modern refineries the bubble caps are replaced by jet trays which are metal sheets with depressions in them.

There is a temperature gradient in the fractionating tower. The top is cooler than the base. Near the base of the tower, heavier hydrocarbons with higher boiling points condense. The lighter hydrocarbons have lower boiling points and so move further up the tower. As they move up the tower, each hydrocarbon condenses at the point where the temperature in the tower falls just below the boiling point of the hydrocarbon. The tower allows the ascending vapour to come into contact with the descending liquid and separation occurs through successive liquid-vapour equilibria. The theory behind this is given in Section 10.5.

The trays and bubble caps allow better mixing and separation of the components.

After a component condenses on a particular tray it moves down to the tray below. When the ascending vapour reaches a tray containing liquid whose temperature is below the boiling point of the vapour, the vapour starts to condense. As it condenses, the vapour heats the liquid in the tray and the more volatile components in the liquid evaporate. The more volatile components pass up the tower with the remaining vapour.

This process occurs in each tray, the least volatile vapour condensing and the most volatile evaporating. The result is that each tray has a fraction containing components with a narrow range of boiling points. So components with lower relative molecular mass products move up the tower and components with higher relative molecular mass move down.

Separating the residue

The residue from the crude oil which has not passed up tower is distilled under vacuum. Lowering the pressure reduces the boiling point and ensures the components distil below their decomposition temperatures (see Section 10.6).

Did you know?

Crude oil has been known about for thousands of years. Oily surface deposits have been written about in ancient Persian tablets. The rich used this crude oil for lighting and for medical purposes. About 1600 years ago, the Chinese were collecting crude oil using bamboo pipes.

☑ *Exam tips*

Make sure that you don't get confused between the words *petroleum* and *petrol*. Petroleum is another name for crude oil. Petrol is a fraction of petroleum used as a fuel in cars. It is better to remember it by its fraction name as gasoline.

Key points

- Fractional distillation of crude oil separates the components into fractions having distinct ranges of boiling points.

- The main fractions obtained by fractional distillation are petroleum gases, gasoline, naphtha, kerosene and fuel oil.

- The petroleum residue is distilled under reduced pressure to form the lubricating oil and bitumen fractions.

- Fractional distillation separates the crude oil into fractions by a series of gas-vapour equilibria which are adjusted as the temperature decreases up the column.

Learning outcomes

On completion of this section, you should be able to:

- describe catalytic cracking and reforming
- describe the impact of the petroleum industry on the environment.

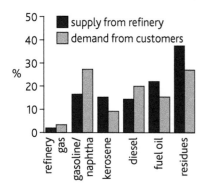

Figure 12.2.1 *Demand and supply for some petroleum fractions*

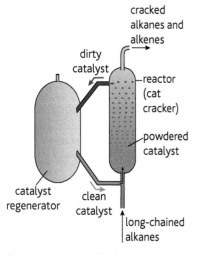

Figure 12.2.2 *Simplified diagram of a catalytic cracking unit*

Cracking

Most of the fractions we get from the distillation of petroleum are useful. But some are more useful than others because there is a greater demand for them. We use more gasoline (petrol) and diesel than can be supplied from the fractional distillation of petroleum (see Figure 12.2.1).

Petroleum companies solve this problem by breaking down larger, less useful hydrocarbons, to smaller, more useful hydrocarbons. They do this by a process called **cracking**. Cracking is the thermal decomposition of alkanes.

When large alkane molecules are cracked, smaller alkane molecules and alkene molecules are formed. Two of the many possible ways that an alkane can undergo cracking are:

$$C_{10}H_{22} \rightarrow C_6H_{14} + C_4H_8$$
$$\text{dodecane} \quad \text{hexane} \quad \text{butene}$$

$$C_{10}H_{22} \rightarrow C_5H_{12} + C_2H_4 + C_3H_6$$
$$\text{dodecane} \quad \text{pentane} \quad \text{ethene} \quad \text{propene}$$

The alkanes formed, e.g. C_6H_{14} are used to make more gasoline. The alkenes are used in chemical synthesis. They are very reactive because of their double bonds. They are the starting compounds (feedstock) for making many new products including plastics and ethanol. Hydrogen (used as a fuel or for making ammonia) can also be produced by cracking ethane.

$$C_2H_6 \rightarrow C_2H_4 + H_2$$
$$\text{ethane} \quad \text{ethene} \quad \text{hydrogen}$$

How is cracking carried out?

Naphtha or gas oil from fractional distillation in an oil refinery (see Figure 12.1.1, p. 134) are the feedstocks. Cracking is usually carried out using a catalyst in a catalytic cracking unit (cat cracker). This type of cracking is called **catalytic cracking**. The vapours from the gas-oil or kerosene fractions are passed through a catalyst of silicon(IV) oxide and aluminium oxide at 400–500 °C. The catalyst is a fine powder and has to be continuously recycled to the cat cracker through a regenerator tank. This frees the catalyst from any carbon deposited on its surface. Modern catalysts include compounds called zeolites which are aluminosilicates.

The main products of catalytic cracking are:

- Refinery gas containing ethene and propene.
- A liquid with a high yield of branched-chain alkanes, cycloalkanes and aromatic hydrocarbons. This is used to make more petrol (especially higher grade petrol).
- A high-boiling point residue. This is used as fuel oil for ships and home heating.

Long-chain alkanes can also be cracked at a high temperature (between 450 and 800 °C). This type of cracking produces a larger percentage of alkenes and is called thermal cracking.

Reforming

Reforming is the conversion of alkanes to cycloalkanes or cycloalkanes to arenes.

When the gasoline and naphtha fractions are passed over a catalyst above 500 °C, the straight-chain alkanes are converted to ring compounds. This process is called cyclisation. See Figure 12.2.3.

The catalyst is platinum or molybdenum(VI) oxide deposited onto aluminium oxide. The Pt or MoO_3 catalyses the dehydrogenation while the aluminium oxide catalyses any rearrangement of the carbon skeleton. More modern plants use bimetallic metal clusters between 1–5 nm in diameter deposited onto aluminium oxide.

A catalyst containing platinum and iridium atoms converts straight-chain alkanes to arenes.

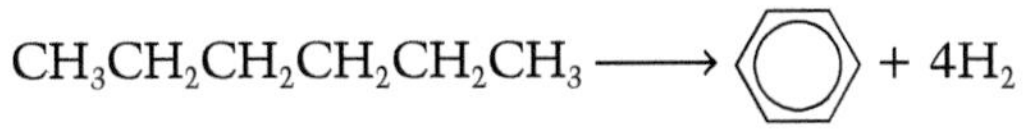

$$CH_3CH_2CH_2CH_2CH_2CH_3 \longrightarrow \bigcirc + 4H_2$$

A catalyst containing platinum and rhenium atoms remove hydrogen from methylcyclohexane to form methylbenzene.

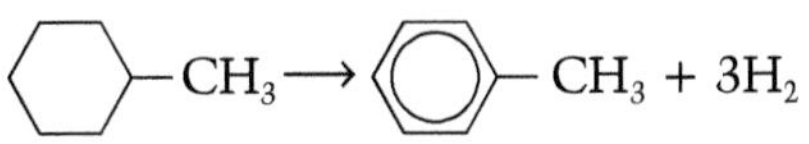

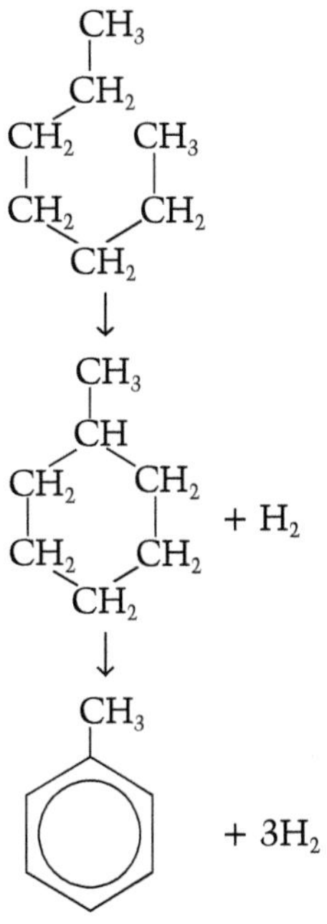

$$\bigcirc\!\!-CH_3 \rightarrow \bigcirc\!\!- CH_3 + 3H_2$$

Petroleum industry and the environment

Crude oil and its refined products are responsible for various types of pollution.

Oil spills

Oil spillages from oil wells or tankers can kill wildlife, especially sea birds and fish. Tar on the birds' feathers reduces their ability to fly and reduces their insulation and ability to float on water. Even a thin layer of oil on the sea results in a large reduction of oxygen in the water underneath, so fish die. Birds, fish and other animals also die through ingesting the toxic components.

Incomplete combustion

Incomplete combustion of petroleum products results in toxic carbon monoxide being formed as well as carbon particles and unburnt hydrocarbons. The latter two can contribute to photochemical smog (see Section 14.8).

Lead

Lead compounds from the addition of tetraethyl lead(IV) as an 'antiknock' agent in gasoline (petrol) can result in damage to the nervous system in children. Although most gasoline does not now contain lead compounds, many people are worried that the arenes put into petrol to replace it are poisonous.

Acid rain

The sulphur present in trace amounts in fuels reacts with oxygen and water in the air to form acid rain. The nitrogen oxides formed when fossil fuels are burnt in vehicle engines also contribute to acid rain (see Section 14.7).

Plastic

Plastics made from petroleum products cause problems in terms of their disposal in the environment and their effect on wildlife (see Section 14.11).

Metals

Some of the metals used as catalysts in the petroleum industry can escape into the air during catalyst change.

Figure 12.2.3 Cyclisation

Did you know?

The worst oil spill from a drilling rig took place in April 2010 in the Gulf of Mexico when a drilling rig exploded. The oil gushed out for three months. Tens of thousands of sea birds and fish died, fishermen were taken ill and many lost their jobs. So large was the oil slick that countries in the Caribbean were on high alert in case the oil reached the region.

Key points

- Cracking of less useful oil fractions produces more useful alkanes and alkenes. The alkanes are used to make petrol and the alkenes to make a wide range of products including plastics.

- Reforming converts straight-chain hydrocarbons to cycloalkanes or arenes.

- Crude oil and its refined products can be responsible for various types of pollution due to accidents during extraction or transport or as a result of combustion of fuels.

13 The chemical industry

13.1 Ammonia synthesis

The Haber Process

Ammonia is manufactured by the Haber Process. The stages in this process are:

- A mixture of nitrogen (1 volume) and hydrogen (3 volumes) is compressed.
- The compressed gases pass into a converter (reactor vessel), which contains trays of catalyst:
  - The catalyst is iron (Fe) or a mixture of iron and iron(III) oxide (the oxide gets reduced by the hydrogen to iron). The iron is porous, so it has a large surface area for the gases to react on. A promoter (usually potassium hydroxide) is added to increase the effectiveness of the catalyst.
  - The temperature in the converter is usually about 400–450 °C
  - The pressure in the converter can range from 25–200 atmospheres (but 200 atmospheres is common).

Under these conditions up to 15% of the nitrogen and hydrogen are converted to ammonia:

- $N_2(g) + 3H_2(g) \rightleftharpoons 2NH_3(g) \qquad \Delta H^{\ominus} = -92 \, kJ \, mol^{-1}$
- The mixture passes into an expansion chamber. The ammonia cools here and condenses. The ammonia is removed as a liquid.
- The unreacted nitrogen and hydrogen are returned to the converter so they are not wasted.

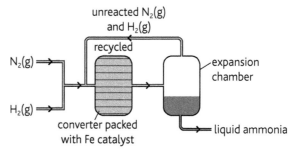

Figure 13.1.1 An outline of the Haber Process

The raw materials for the Haber Process

The hydrogen is made either from natural gas or by cracking ethane obtained from the fractional distillation of crude oil. It can also be made from natural gas by reaction with steam in the presence of a nickel catalyst.

$$CH_4(g) + H_2O(g) \xrightarrow{\text{heat + Ni}} CO(g) + 3H_2(g)$$

The carbon monoxide which can poison the catalyst used in the Haber Process is removed by reaction with more steam.

$$CO(g) + H_2O(g) \rightarrow CO_2(g) + H_2(g)$$

The nitrogen for the Haber Process is extracted from the air (by fractional distillation of air). The oxygen from the air is removed by reaction with hydrogen.

The best conditions for the Haber Process

Effect of pressure

Ammonia production is favoured by an increase in pressure. An increase in pressure shifts the equilibrium towards the right. More product is formed. This is because according to Le Chatelier's Principle (see Book 1, Section 8.5) increasing the pressure shifts the equilibrium in favour of fewer gaseous molecules. A pressure between 25 and 200 atm is used, depending on the plant. Although pressures above 200 atmospheres give a higher yield they are not used because:

- A lot more energy is required to power the compressors. This costs a lot more money.
- At higher pressures the reaction vessels are less safe. A lot more money would have to be spent to make them strong enough to withstand the extra pressure.

Effect of temperature

Ammonia production is favoured by lower temperature. This is because the reaction is exothermic. For an exothermic reaction an increase in temperature decreases the value of K_p so decreases the yield of the forward reaction, i.e. the yield of ammonia. This is because according to Le Chatelier's principle:

- decrease in temperature decreases the energy of the surroundings
- the reaction goes in the direction in which energy is released
- energy is released in the exothermic reaction. This favours the reactants.

Effect of catalyst

A catalyst does not affect the yield of ammonia but does increase the rate at which the product (ammonia) is formed.

The best conditions overall

Figure 13.1.2 shows how the yield varies with temperature and pressure.

When the temperature is increased:

- the rate of reaction increases
- the equilibrium yield decreases.

There is a conflict between the between the best equilibrium yield, which decreases with increase in temperature and the best rate of reaction which increases with increase in temperature. So we use compromise conditions; a temperature of about 420–450 °C is used at 200 atmospheres with an iron catalyst to give a reasonable yield at a fast enough rate.

Removing ammonia by condensing it also helps improve the yield. This is because removing ammonia as a liquid shifts the position of equilibrium to the right in favour of fewer molecules.

Did you know?

The Haber Process is named after the German chemist Fritz Haber, who first discovered this process in the early twentieth century. The German chemist Carl Bosch improved the process after testing thousands of catalysts in order to find the best ones. As a result the process's full name is the Haber–Bosch Process.

Haber and Bosch were awarded Nobel prizes for their work.

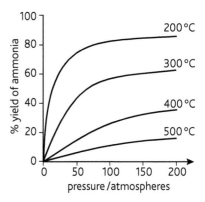

Figure 13.1.2 *The yield of ammonia depends on both the temperature and the pressure*

Key points

- Ammonia is manufactured by the Haber Process:

$$N_2(g) + 3H_2(g) \rightleftharpoons 2NH_3(g)$$

- The conditions of the Haber Process are 200 atm, 450 °C and Fe catalyst.
- The yield of ammonia decreases as the temperature increases.
- The rate of production of ammonia increases as temperature increases.
- The conditions used in the Haber Process are a compromise between a high equilibrium yield and high rate of reaction.

The uses of ammonia

Ammonia is made on a huge scale and has many uses (see Figure 13.2.1). About 85% of the ammonia produced is used to make fertilisers.

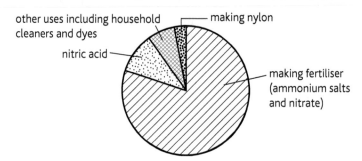

Figure 13.2.1 *The main uses of ammonia*

Apart from making fertilisers, the main uses of ammonia are:

- Making other nitrogen-containing compounds, especially nitric acid. Nitric acid is used to make fertilisers, explosives and many organic nitrogen compounds in the pharmaceutical and dye industries.
- As components of household cleaners for shiny surfaces such as glass and ovens. Ammonia cleans without leaving 'streaky' marks.
- As a refrigerant, especially on a large scale, e.g. for cooling ice rinks.
- In fermentations as a source of nitrogen for microorganisms and to adjust the pH of the fermentation mixture.
- In treating textiles (especially cotton and wool) to alter their properties.
- To make dyes.

Manufacturing fertilisers

Plant roots absorb nitrogen from the soil in the form of nitrates. The plants convert the nitrates to proteins which are needed for growth. When farmers harvest their crops, the nitrogen taken up by the plant is not usually returned to the soil. Over a number of years, the soil becomes depleted in nitrates. Unless nitrates (and other minerals) are put back into soil, future crops will not grow as well as they should. In the modern agricultural industry, natural fertilisers such as manure are not available in sufficient quantities to add nitrogen back to the soil so fertilisers are used to provide nitrates and other essential minerals for plants to grow. When fertilisers are used to provide more nitrogen, plants grow faster and bigger. This increases the crop yield.

Ammonia can be used as a fertiliser (mainly in the USA) but it evaporates readily from the soil. So ammonium compounds such as ammonium nitrate, ammonium sulphate and ammonium phosphate are used. These are made from ammonia and the relevant acid.

e.g.

$$NH_3(aq) + HNO_3(aq) \rightarrow NH_4NO_3(aq)$$
ammonia   nitric acid   ammonium nitrate

$$3NH_3(aq) + H_3PO_4(aq) \rightarrow (NH_4)_3PO_4(aq)$$
ammonia   phosphoric acid   ammonium phosphate

☑ Exam tips

When writing equations for the formation of ammonium salts from ammonia, remember that no water is formed as a product. e.g.

$$2NH_3(aq) + H_2SO_4(aq) \rightarrow (NH_4)_2SO_4(aq)$$

The solutions are evaporated and sprayed into a tower into which air is blown. Hard pellets of the fertiliser are formed. These dissolve slowly in the soil. Fertiliser factories often have several chemical plants next to each other making:

- ammonia by the Haber Process
- nitric acid from ammonia
- nitrate or phosphate fertilisers from the ammonia and nitric acid.

Most fertilisers are called NPK fertilisers because they contain nitrogen, phosphorus and potassium, all three of which are needed for healthy plant growth.

Ammonia and the environment

Eutrophication

Overuse of fertilisers causes **eutrophication**. This is process by which excess quantities of fertilisers pollute rivers and lakes and cause an overgrowth of algae and bacteria leading to the death of aquatic organisms. The stages are:

- Rainwater dissolves fertilisers and the solution runs off (leaches) from fields into rivers and lakes.
- The concentration of nitrates and phosphates in the river or lake increases.
- Algae in the water use these nutrients. They grow very fast causing an algal bloom covering the surface of the water.
- The dense growth of algae blocks sunlight from reaching plants below the water surface.
- These plants die from lack of sunlight. The algae also die when the nutrients are used up.
- Bacteria feed on the dead plants and algae and multiply rapidly.
- The bacteria use up the oxygen dissolved in the water.
- Without oxygen, fish and other water animals die.

Other effects of ammonia

Smog: Ammonia in the atmosphere can combine with nitrogen and sulphur oxides from vehicles and industry to form fine particles which contribute to smog.

Human health: Ammonia itself can irritate the lungs and inhibit the uptake of oxygen by haemoglobin by altering the pH of the blood. Ammonia can react with acids in the atmosphere to form ammonium salts. These exist as small particles (particulates). When breathed in over a period of time, these can cause bronchitis, asthma, coughing fits and 'farmer's lung'.

Soil acidification: When ammonia in the atmosphere reacts with water in the soil it is converted to NH_4^+ ions. NH_4^+ ions are also present in fertilisers. Excess NH_4^+ ions are converted by bacteria to nitrites, nitrates and H^+ ions. The H^+ ions make the soil acidic and plants may not be able to grow well.

Changes to plant diversity: Ammonia gas can settle on plant leaves and stems and cause damage because of its alkalinity especially in alpine plants.

Did you know?

Trinidad is the largest exporter of fertilisers in the Caribbean and is one of the world's largest exporters of fertilisers. The 3-plant ammonia and fertiliser factory at Savonetta in central Trinidad exports nearly 99% of its production.

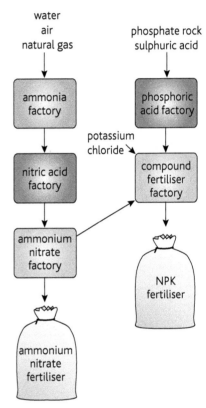

Figure 13.2.2 *A flow chart for making NPK fertilisers*

Key points

- Most ammonia is used to make fertilisers.
- Fertilisers are added to the soil to replace the nitrogen taken up by crop plants and to increase crop yield.
- The main impact of fertilisers on the environment is eutrophication.
- Ammonia can react with acids in the air to form particulates which can be damaging to health.

On completion of this section, you should be able to:

- explain the importance of fermentation and distillation in the manufacture of alcoholic beverages

- describe the uses of ethanol as a fuel and in the pharmaceutical industry.

The production of alcoholic beverages

Fermentation

The drinks industry is big business in many parts of the world. For example in Jamaica, the rum industry is worth about 45 million dollars in exports. Alcoholic drinks include beer, wine and spirits such as rum, whisky and gin. **Fermentation** has been used by humans for centuries to produce alcohol. Fermentation is the conversion of carbohydrates to alcohols or organic acids using yeast or bacteria in the absence of air. Almost any vegetable material can be used as starting materials as long as they contain sugar or starch which can be broken down to simple sugars such as glucose. In the production of alcoholic drinks, yeast acts on glucose or sucrose in the absence of air to produce CO_2 and ethanol.

From glucose: $C_6H_{12}O_6(aq) \rightarrow 2C_2H_5OH(aq) + 2CO_2(g)$

From sucrose: $C_{12}H_{22}O_{11}(aq) + H_2O(l) \rightarrow 4C_2H_5OH(aq) + 4CO_2(g)$

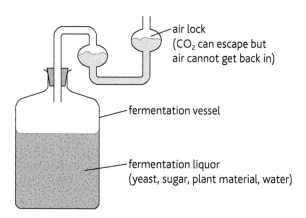

air lock
(CO_2 can escape but air cannot get back in)

fermentation vessel

fermentation liquor
(yeast, sugar, plant material, water)

Figure 13.3.1 *Simple fermentation apparatus*

Glucose is mixed with yeast and water and left at a temperature of between 15 and 40 °C (according to the type of alcoholic beverage required). Temperature control is important:

- Too low a temperature will cause the enzymes to work too slowly and increases the possibility of unwanted bacterial fermentation.

- Too high a temperature will alter the cell structure of the yeast and proteins will have the incorrect structure to catalyse the reaction. At higher temperatures, the proteins will become denatured.

It is important to keep air out of apparatus to prevent:

- oxidation of ethanol to ethanoic acid

- unwanted side reactions due to bacterial action in the presence of air.

Fermented drinks are generally classified according to the plant material used in the fermentation. For example:

- Beers are made from cereals and other starchy material.

- Wines and ciders are made from fruit juices.

- Mead is made from honey.

Distillation

There is a limit to the amount of alcohol that can be produced by fermentation. This is because when the alcohol content rises to more than 15% it will kill the yeast. Spirits are made by distilling the fermentation mixture when it reaches a suitable alcoholic content. The alcohol content in spirits is much higher than in wines or beers, e.g. 35–40% by volume ethanol. In the distillation of the fermented liquor to make rum, some producers work in batches (pot stills). Heat is applied directly to the pot still and the alcohol and aroma compounds giving the rum its characteristic flavours evaporate.

Other producers use column distillation. See Section 10.9 for further information.

Two important spirits are:

- Rum from distilling fermented molasses or sugar cane juice.
- Whisky from fermented cereal grains.

Uses of ethanol

Ethanol as a fuel

The main use of ethanol is as a fuel or a fuel additive for vehicles. Brazil has the largest ethanol fuel industry in the world. It is produced by fermentation of sugar cane residues. In the USA, ethanol is largely made from corn (maize).

Ethanol can either be used on its own as a fuel or mixed with gasoline.

- Ethanol containing 4.9% water is produced by distillation. This is used for vehicles that run on ethanol only.
- Water is removed by an adsorbent, such as a zeolite or starch, if ethanol for mixed petrol-ethanol engines is required.

Other uses of ethanol

- Alcoholic beverages (see above).
- In the chemical industry to make halogenoalkanes, esters, ethers, ethanoic acid and amines.
- Ethanol is used (usually as methylated spirits) as an industrial solvent in paints and adhesives.
- It is a commonly used solvent in perfume industry because it evaporates rapidly.
- As an antiseptic: it is used in medical wipes and antibacterial hand gels.
- As an antidote to poisoning by other alcohols, e.g. ethylene glycol poisoning.

Did you know?

It is thought that the word *alcohol* first came into the English language in 1543 from an Arabic word 'al-ghul'.

Key points

- Fermentation is used to produce alcoholic drinks.
- In order to make alcoholic drinks such as rum and whisky, the fermentation mixture is distilled.
- Ethanol from the fermentation of sugar cane residues or other plant material can be used as a fuel.
- Ethanol has other uses including as a solvent, as an antiseptic and in perfume manufacture.

Learning outcomes

On completion of this section, you should be able to:

- describe the social and economic impact of ethanol production and consumption
- describe the physiological changes caused by alcohol consumption
- describe the impact of the ethanol industry on the environment.

Did you know?

The Latin saying 'in vino veritas' (in wine, the truth) suggests that people are more likely to tell the truth when they have had a glass of wine or two to drink!

Did you know?

When used as a solvent about 10% methanol is often added to ethanol to 'denature' the alcohol so that it is not fit for drinking. This alcohol mixture is called methylated spirits. Some alcoholics, however, who cannot afford alcoholic drinks, resort to drinking methylated spirits. This can result in blindness and eventually death. Because of this, some manufacturers of methylated spirits now add emetics to the methylated spirits to make people vomit up the alcohol.

The effect of ethanol on the body

Although the ethanol in alcoholic beverages tends to make people relax on social occasions, ethanol is classed as a psychoactive drug (one which acts on the central nervous system and results in changes in mood, how we view things, understanding, and behaviour). If we drink alcohol in small amounts we may get a feeling of general well-being and relaxation. As the amount of alcohol in the bloodstream increases (measured as blood alcohol content, BAC, in $g\,dm^{-3}$), the ethanol has a progressively bad effect on us. The short term effects are:

- BAC $0.5\,g\,dm^{-3}$: feeling of relaxation, increased talkativeness, impaired judgement.
- BAC $1.0\,g\,dm^{-3}$: difficulty moving properly, giddiness, feeling of not being in control, nausea, vomiting, symptoms of intoxication, e.g. slurred speech, aggressive behaviour.
- BAC $3.0\,g\,dm^{-3}$: Not knowing what is happening to oneself, loss of consciousness.

At blood alcohol concentrations of above about $1\,g\,dm^{-3}$, ethanol acts as a depressant, it lowers the activity of particular parts of the brain. At blood alcohol concentrations above $1.4\,g\,dm^{-3}$ ethanol decreases the flow of blood to the brain, leading to loss of consciousness and eventually possible death at concentrations above $4\,g\,dm^{-3}$.

Long-term effects of excessive consumption of alcohol (alcoholism) can lead to many social and health problems. Although small amounts of ethanol can be metabolised in the body and can be used as a source of energy, one product of the alcohol metabolism is ethanal and this is carcinogenic (causes cancer). A buildup of such toxic substances in the body may result in liver cancer.

Excessive alcohol consumption in pregnant women can result in foetal alcohol syndrome in which the child of the alcoholic mother can have various birth defects.

The social and economic impact of ethanol

These include:

- Social consequences for individuals injured or for the families of those killed as a result of the impaired judgement of a driver under the influence of drink.
- Long-term isolation of individuals who have an alcohol habit and who may require more help from the community.
- More treatment in hospitals required. In many countries in the world the number of expensive liver transplants required due to long-term alcohol consumption is increasing rapidly.
- Many working hours are lost when people do not turn up for work because of a 'hangover'. This has an economic impact on production if there is nobody to take their place.

Ethanol production and the environment

When ethanol is made by fermentation:

- It helps conserve the world's diminishing supply of crude oil as it is used as an alternative fuel.

- The process is 'carbon neutral'. The sugar cane absorbs CO_2 as it grows and, although this CO_2 is returned to the atmosphere when ethanol burns, the two processes are in balance. (Burning fossil fuels just puts CO_2 into the atmosphere, see Section 14.4.)

- CO_2 (a greenhouse gas) is produced when harvesting, transporting and processing the sugar cane and distilling it. But this is still cheaper and better for the environment than extracting petroleum from the ground and fractionally distilling it.

- When combusted in the car engine, ethanol produces greenhouse gases. But the amounts formed are less than the amounts formed in an engine running on gasoline (petrol) or diesel. It also produces relatively less carbon monoxide and particulates in comparison with petrol or diesel derived from crude oil.

- The combustion of ethanol in car engines produces approximately twice as much methanal and ethanal compared with a petrol or diesel engine. These aldehydes undergo photochemical activity in the atmosphere generating low level ozone, so leading to smog.

- Food production might decrease or get more expensive. The production of biofuels such as ethanol has increased in recent years. Some crop plants such as corn (maize), sugar cane and vegetable oils can be used either as food, animal feed or to make biofuels. If crop plants are grown for biofuels instead of for food, food will become more expensive or in some areas scarcer.

- The increased production of ethanol may lead to more land being cleared to grow sugar cane or other plants being grown. The increased destruction of forests or other areas of countryside is likely to:
  - destroy animal and plant habitats and lead to species' extinction
  - displace people living in these areas
  - decrease the carbon sinks (see Section 14.4) and cause soil erosion.

Comparing methods

There are two main ways of producing ethanol:

- fermentation
- hydration of ethane $CH_2{=}CH_2(g) + H_2O(g) \rightarrow CH_3CH_2OH(l)$

When comparing these methods and their effect on the environment we have to consider the following:

Fermentation	Hydration of ethene
Easy to set up	Complex to set up
Requires low temperatures, e.g. 15–40 °C	Requires temperature of 300 °C
Requires natural catalyst (yeast)	Requires phosphoric acid catalyst
Produces ethanol of about 15% concentration so distillation required	Produces very pure ethanol
Raw materials are plants and yeast	Raw materials from distillation of petroleum

You can see that the energy requirements for the production of ethanol by hydration of ethene are likely to be higher than for fermentation and more CO_2 (greenhouse gas) is likely to be emitted.

Did you know?

Ethanol can also be made from cellulose in straw, sawdust and other wood material. Enzymes which digest cellulose have been used to produce sugars which can then be fermented.

Key points

- Ethanol is a psychoactive drug which depresses the nervous system.
- Drinking alcohol leads to feeling of giddiness, feeling of not being in control, nausea, vomiting and other symptoms of intoxication.
- Drinking alcohol to excess over a long period can lead to death.
- The use of ethanol as a fuel rather than gasoline or diesel is better for the environment as less CO_2 is produced in making ethanol by fermentation than in making petrol or diesel.

On completion of this section, you should be able to:

- describe the chemical processes involved in the electrolysis of brine using the diaphragm cell

- describe the economic advantages of chlorine production by the diaphragm cell method

- describe the production of sodium hydroxide by the electrolysis of brine.

The electrolysis of brine

The diaphragm cell

Brine is a concentrated aqueous solution of sodium chloride. It is obtained from seawater or by dissolving rock salt in water. The electrolysis of brine is used to produced chlorine, hydrogen and sodium hydroxide. Fig 13.5.1 shows a diaphragm cell used to electrolyse brine.

- The cell is divided into a series of cathode and anode compartments.
- The electrolyte is a concentrated solution of brine (sodium chloride).
- The anodes are titanium rods.
- The cathodes are steel grids.
- A porous diaphragm separates the cathode and anode compartments. This is made of a mixture of asbestos and polymers. Water and ions can pass through the diaphragm.

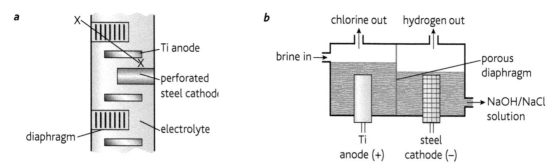

Figure 13.5.1 *A diaphragm cell;* **a** *The arrangement of the electrodes from above;* **b** *A simplified diagram of the cell across X-X in* **a**

The ions present in the electrolyte are:

- Sodium, Na^+
- Chloride, Cl^-
- Hydrogen ions, H^+, from the self ionisation of water
- Hydroxide ions, OH^-, from the self ionisation of water.

$$H_2O(l) \rightleftharpoons H^+(aq) + OH^-(aq)$$

The electrode reactions

At the anode

Both Cl^- and OH^- ions move to the anode. Only Cl^- ions undergo oxidation. This is because they are in far greater concentration than the OH^- ions. The chlorine gas is pumped off from the top of the anode compartment.

$$2Cl^-(aq) \rightarrow Cl_2(g) + 2e^-$$

At the cathode

Both Na^+ and H^+ ions move to cathode. Only H^+ ions undergo reduction. This is because hydrogen is lower in the discharge series (and the electrochemical series) than sodium. The hydrogen gas is pumped off from the top of the cathode compartment.

$$2H^+(aq) + 2e^- \rightarrow H_2(g)$$

The Na^+ ions remain in the cathode compartment.

Formation of sodium hydroxide

The removal of H^+ ions causes the following equilibrium to shift to the right.

$$H_2O(l) \rightleftharpoons H^+(aq) + OH^-(aq)$$

As more and more H^+ ions are removed, the concentration of OH^- ions in the cathode compartment increases. So Na^+ and OH^- ions, the components of sodium hydroxide are present. The electrolyte level in the anode compartment is kept higher than in the cathode compartment. This ensures that the flow of electrolyte is towards the cathode compartment and so reduces the possibility of NaOH moving to cathode compartment.

When concentrated enough, the solution containing 10% NaOH and 15% NaCl by mass is run off from the cathode compartment. This solution is partially evaporated and the NaCl removed by crystallisation leaving the more soluble NaOH as a 50% weight/ volume solution.

The economic advantages of a diaphragm cell

A lot of sodium hydroxide and chlorine is still manufactured using a mercury cell (Castner cell) see 'Did you know?' box.

Did you know?

In the mercury cell method, purified brine flows through the cell in the same direction as the mercury. Cl_2 is formed at the Ti anode. At the mercury cathode Na^+ ions are discharged in preference to H^+ ions because of a high overvoltage. The mercury/ sodium mixture (amalgam) is then sent to an amalgam decomposer, where the sodium reacts with water to form a solution of sodium hydroxide.

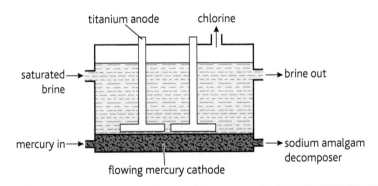

The diaphragm cell has several advantages over the mercury cell. The advantages and disadvantages are summarised in the table.

Key points

- The diaphragm cell has Ti anodes and steel cathodes and an electrolyte of brine (concentrated aqueous NaCl).
- In the diaphragm cell Cl_2 is formed at the anode and H_2 at the cathode. The solution in the cathode compartment is a solution of NaOH and NaCl from which the NaOH is separated.
- The diaphragm cell works at a lower voltage than the mercury cell and does not contain toxic mercury.

Did you know?

A more modern way of making chlorine and sodium chloride is to use a special ion-permeable membrane rather than a diaphragm. Membrane cells produce a higher concentration of sodium hydroxide than diaphragm cells so costs in concentrating the NaCl/NaOH mixture are reduced. The possibility of H_2 and Cl_2 mixing is also reduced and the membranes are longer lasting than the diaphragms.

✅ *Exam tip*

You do not have to know the details of the mercury cell or about overvoltage. You should concentrate on the economic advantages of the diaphragm cell over the mercury cell.

Mercury cell	Diaphragm cell
Expensive to construct	Much cheaper to construct
Works at 4.5V (more expensive to run)	Works at 3.8V (slightly less expensive to run)
Toxic waste mercury must be removed	No toxic mercury
No asbestos diaphragm	Asbestos diaphragm needs to be renewed quite often and asbestos dust is toxic
Sodium hydroxide purer	Sodium hydroxide less pure
Needs high purity brine to work	Works with brine of fairly low purity

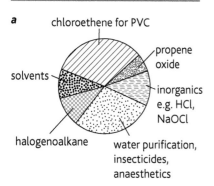

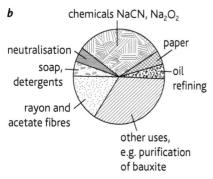

Fig 13.6.1 *Uses of* **a** *chlorine and* **b** *sodium hydroxide*

The importance of the chlor-alkali industry

The manufacture of chlorine and sodium hydroxide together is known as the chlor-alkali industry. Both chlorine and sodium hydroxide are the starting materials for many chemical processes which produce plastics, solvents and aerosols amongst other things. There are about 15 000 different chlorine-containing compounds which are used commercially! At the end of the last century, there was a huge growth in the demand for chlorine due to the increased use of chloro-organic compounds such as solvents and plastics. Before this there was always an excess of NaOH produced by the mercury diaphragm cells. The uses of chlorine and sodium hydroxide produced by the chlor-alkali industry are shown in Figure 13.6.1.

The use of chlorine in making solvents and halogenoalkanes is now decreasing because of the toxicity of the products made and environmental problems (see below).

Uses of the halogens

Fluorine

- To make uranium hexafluoride for the production of nuclear 'fuel'.
- To make sulphur hexafluoride, which is an inert medium for some electrical work.
- To make PTFE, which is a 'non-stick' plastic for cooking pans, etc.
- To make hydrofluoric acid for etching glass.
- To make hydrofluorocarbons for anaesthetics.

Chlorine

- To make bleaches, which often contain sodium chlorate(I) (sodium hypochlorite).
- To make vinyl chloride, the monomer for the plastic, polyvinyl chloride (PVC).
- To make halogenoalkanes for chemical syntheses.
- To make anaesthetics (which often have fluorine in them as well).
- To make aerosols (although the production of chlorine-containing aerosols has decreased recently due to their effect on the ozone layer).
- To make solvents such as trichloroethane (the production of these is also decreasing due to their toxicity as well as their effect on the ozone layer). Some of these solvents are still used in dry cleaning.
- To make refrigerants. Chlorine- and fluorine-containing hydrocarbons (chlorofluorocarbons) are good refrigerants but their use has declined in recent years (see Section 14.3).
- To make insecticides and dyestuffs.
- Sterilisation in swimming pools and water treatment works.

Bromine

- Polymers containing bromine atoms are good flame retardants.
- Making pesticides, dyes and some pharmaceutical products.
- Making bromoethene as an antiknock agent to allow gasoline (petrol) to combust properly in a car engine (although becoming rarer).

Iodine

- As a catalyst in the production of ethanoic acid by the Monsanto process.
- As an additive to the feed for cows, sheep and pigs (nutrient supplement).
- To disinfect water and in water treatment.
- Often added to table salt (to help prevent the disease called goitre).

The chlor-alkali industry and the environment

Chlorine-containing compounds as well as the method of production of chlorine may have an impact of the environment.

Mercury waste

Mercury is toxic. About one-third of the chlorine produced is made using the Castner cell which has a flowing mercury cathode (see Section 13.5). Mercury and mercury compounds formed during the operation of the cell can escape into the air or water. Even small amounts of mercury can kill fish and poison people who eat the fish.

Asbestos

The asbestos used in the diaphragm of the diaphragm cell (see Section 13.5) has to be changed regularly. When it dries out, asbestos fibres are released into the air. Tiny amounts of these fibres can cause the lung condition 'asbestosis' in which breathing becomes very difficult and there is an increased risk of lung cancer.

Sodium hydroxide leaks

Sodium hydroxide can occasionally leak from the cell or get into the environment from the evaporation process. Sodium hydroxide is a strong base and hence can alter the pH of water sufficiently to cause death to some organisms in the water.

Depletion of the ozone layer

Many of the chlorofluorocarbons, tetrachloromethane and 1,1,1-trichloroethene used as solvents, aerosols and refrigerants were banned in 1997 because they are ozone-destroying chemicals. Other halogenoalkanes may also destroy ozone. (For details see Section 14.3).

Polyvinyl chloride (PVC)

This plastic is used in huge amounts in the construction, packaging and other industries. PVC is not biodegradable so waste PVC contributes to litter and landfill (see Section 14.11). PVC is sometimes burnt in controlled waste recycling and as a source of energy. These processes can put poisonous dioxins and acidic hydrogen chloride into the atmosphere.

Dioxin emissions

The effluents from the paper industry, which can use chlorine as a bleaching agent, may contain halogenocarbon compounds (which may cause ozone depletion) and dioxins (which are very poisonous).

Did you know?

Salt (sodium chloride) was very important to the Romans for preserving food and tanning leather. The Romans sometimes gave their soldiers money to buy salt. The word salary comes from the latin 'sal' meaning 'salt' and 'arius' meaning 'connected with'.

Did you know?

'Minamata disease' (a form of mercury poisoning) was named after the town of Minamata in Japan. Mercury-containing waste from a factory which made ethanal had been seeping into the water in Minamata Bay since 1932. The chemical was absorbed by fish. In 1956 an 'epidemic of an unknown disease of the central nervous system' was reported. This was caused by people in the area eating fish containing the poisonous mercury.

Key points

- Chlorine is used to make bleaches, PVC, halogenoalkanes, solvents, aerosols, refrigerants and anaesthetics.

- Environmental problems related to the chlor-alkali industry include mercury poisoning, ozone depletion, asbestosis and non-biodegradability of PVC.

The manufacture of sulphuric acid

Most sulphuric acid is made by the **Contact Process**. The raw materials for this process are:

- sulphur (from sulphur deposits beneath the ground, from sulphide ores or from hydrogen sulphide from petroleum or natural gas)
- air (from the atmosphere)
- water.

There are three stages in the process: sulphur burning, sulphur dioxide conversion and absorption. These are shown in Figure 13.7.1.

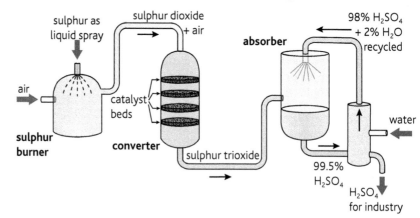

Fig 13.7.1 *The manufacture of sulphuric acid by the Contact Process*

Sulphur burning

A spray of molten sulphur is burned in a furnace in a current of dry air.

$$S(l) + O_2(g) \rightarrow SO_2(g)$$

The gas mixture which comes out of the burner contains about 10% sulphur dioxide and 10% oxygen by volume.

Sulphur dioxide conversion

This is the key reaction in the process. The sulphur dioxide is passed into a reaction vessel (the converter). The converter contains several layers (usually four) of vanadium(v) oxide catalyst, V_2O_5, on a silica support. In the converter sulphur dioxide is converted to sulphur trioxide.

$$2SO_2(g) + O_2(g) \rightleftharpoons 2SO_3(g) \; \Delta H^\ominus = -98\,kJ\,mol^{-1}$$

Since the reaction is exothermic, the heat is removed between each layer of catalyst by heat exchangers. The percentage conversion of SO_2 to SO_3 is between 96–99.5%.

Absorption

The sulphur trioxide is absorbed into a 98% solution of sulphuric acid. This happens in a tower called an absorber. The tower is packed with ceramic material. The sulphur trioxide is not absorbed directly into water. This is because a mist of corrosive sulphuric acid is formed when sulphur trioxide reacts with water and this does not condense very easily. The sulphur trioxide dissolves in the 98% sulphuric acid to form a thick liquid called oleum.

$$SO_3(g) \;+\; H_2SO_4(l) \;\rightarrow\; H_2S_2O_7(l)$$
$$\text{98\% sulphuric acid} \qquad \text{oleum}$$

When oleum is mixed with a little water, 98% sulphuric acid is reformed.

Some of this acid is returned to the absorber. The rest is run off to be used as concentrated sulphuric acid.

$$H_2S_2O_7(l) + H_2O(l) \rightarrow 2H_2SO_4(l)$$

The best conditions for the Contact Process

The key reaction in the Contact Process is:

$$2SO_2(g) + O_2(g) \rightleftharpoons 2SO_3(g) \; \Delta H^{\ominus} = -98\,\text{kJ mol}^{-1}$$

Effect of pressure

Sulphur trioxide production will be favoured by an increase in pressure. When the pressure is increased the position of equilibrium shifts to the right. More product is formed. This is because according to Le Chatelier's principle (see *Unit 1 Study Guide*, Section 8.5) increasing the pressure shifts the equilibrium in favour of fewer gaseous molecules. Most plants, however, operate either at atmospheric pressure or at temperatures only slightly above atmospheric pressure. This is because:

■ The percentage yield of the reaction is very high without increasing the pressure. The marginally increased yield would not compensate for the extra energy requirements needed to produce higher pressure. Only very large scale plants use pressures above atmospheric pressure.

■ At higher pressures, there are additional problems because of the corrosive nature of the moist gases.

Effect of temperature

Sulphur trioxide production is favoured by lower temperature. This is because the reaction is exothermic. For an exothermic reaction an increase in temperature decreases the value of K_p so decreases the yield of the forward reaction, i.e. the yield of sulphur trioxide.

The temperature range in the converter is between 450–580 °C. Because the reaction is exothermic, heat exchangers are used to try to decrease the temperature so that the catalyst is within its working range.

Effect of catalyst

The catalyst does not affect the yield of sulphur trioxide but does increase the rate at which it is formed. The vanadium(v) oxide catalyst is inactive below about 370 °C. It works best at about 410 °C.

The best conditions overall

When the temperature is increased:

■ the rate of reaction increases

■ the equilibrium yield decreases.

Since the yield of sulphur trioxide is high at atmospheric pressure, there is no point in wasting extra energy pressurising the converter. The temperature is maintained at about 450 °C so that there is a good reaction rate in the region where the catalyst is most efficient.

✓ *Exam tips*

In an exam, make sure that you distinguish between a question that asks for the best conditions for converting SO_2 to SO_3 and one which asks for the actual conditions in the Contact Process. The answer to the former is low temperature, high pressure and V_2O_5 catalyst. The answer to the latter is 450 °C, atmospheric pressure and V_2O_5 catalyst.

Key points

■ The key reaction in the Contact Process is

$$2SO_2(g) + O_2(g) \rightleftharpoons 2SO_3(g)$$

■ The conditions used in the Contact Process are 450 °C, atmospheric pressure and catalyst of vanadium(v) oxide.

■ Sulphur trioxide is absorbed in 98% sulphuric acid rather than water to make oleum. More sulphuric acid is then made by add a little water to the oleum.

Sulphur and some compounds of sulphur

About 0.1% of the Earth's crust consists of sulphur. It is found as the element in salt domes in the USA and Mexico and associated with other minerals in various parts of Europe. About 90% of the sulphur mined is used in the production of sulphuric acid. In the production of sulphuric acid, the sulphur is first burnt to make sulphur dioxide (see Section 13.7). The other 10% is used to make chemicals for agriculture, dyestuffs, in woodpulping processes and in vulcanisation.

- Vulcanisation: In the manufacture of tyres, sulphur is added to the rubber to make the tyre harder. The rubber becomes less sticky. An 'accelerator' is also added to speed up the process.
- Sulphur powder: This is used as a fungicide for dusting on plants such as vines and strawberries.
- Carbon disulphide: This used for making the polymers rayon and cellophane.
- Pharmaceuticals: Some drugs and medicines are sulphur compounds, e.g. sulphonamides.
- Organic sulphur compounds: Some dyes and agrochemicals contain sulphur.

The uses of sulphur dioxide

Sulphuric acid manufacture

Sulphur is burnt in air to produce the sulphur dioxide required for the manufacture of sulphuric acid by the Contact Process.

Food preservation

Sulphur dioxide is used to preserve food and drinks. It does this by killing any bacteria present. Sulphur dioxide can be added directly to drinks such as wine and dried fruits such as apricots in order to preserve them. In general, it is sulphites such as sodium sulphite, which are added to foods such as packaged meats and ready-made meals. In acidic conditions, which may be caused by bacterial fermentation, sulphur dioxide is released. The sulphur dioxide kills the bacteria.

$$SO_3^{2-}(aq) \ + \ 2H^+(aq) \ \rightarrow SO_2(g) + H_2O(l)$$
$$\text{sulphite ion} \quad \text{acid (in food)}$$

Because it is a reducing agent, sulphur dioxide also acts as an antioxidant, preventing food and drink reacting with the air.

The bleaching action of sulphur dioxide

Sulphur dioxide is used as a bleach during the manufacture of paper from wood pulp. It is especially useful for bleaching silk, wool and straw which are damaged by stronger bleaches such as chlorine.

Other uses

Sulphur dioxide was formerly used as a refrigerant. In chemistry liquid SO_2 is used as an inert solvent. Sulphur dioxide gas is a good reducing agent.

The uses of sulphuric acid

Figure 13.8.1 shows the main uses of sulphuric acid.

Among other things, sulphuric acid is used:

- to make phosphoric acid. Sulphuric acid is added to calcium fluorophosphate rocks to produce the acid. Phosphoric acid is used to make phosphate fertilisers
- to make ammonium sulphate which is a fertiliser
- as a cleaning agent for metal surfaces
- as the electrolyte in lead/acid car batteries
- as a catalyst in various chemical processes
- to make detergents. Many of these are sulphonates
- to make dyes and paints
- to make corrosion-resistant concrete.

The impact of the sulphuric acid industry

In general sulphuric acid itself, although corrosive, is not toxic. However, the sulphur oxides used in the production of sulphuric acid are toxic as are a number of sulphur compounds such as hydrogen sulphide.

- In the Contact Process (or other processes) used to make sulphuric acid, sulphur dioxide or sulphur trioxide must not be allowed to escape into the atmosphere as they are toxic and cause acid rain (see Section 14.7).
- Sulphuric acid escaping into the environment from various industrial processes may acidify rivers and lakes causing death of animals and plants which are acid sensitive.
- Sulphuric acid and sulphur dioxide can cause acidification of soils. This can result in loss of minerals by leaching.
- The presence of sulphur dioxide in the atmosphere can lead to irritation of the eyes and throat.
- Acid aerosols present in the atmosphere which include gaseous sulphur dioxide can present a fire hazard.
- The contact of metals with a sulphuric acid spill can result in the liberation of hydrogen gas which could explode and cause fires.
- Since sulphuric acid is used to make phosphate fertilisers, the sulphuric acid industry contributes indirectly to eutrophication (see Section 13.2).

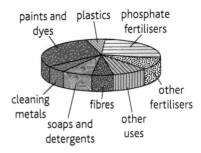

Figure 13.8.1 *The uses of sulphuric acid*

Did you know?

Jābir ibn Hayyān, who is sometimes called 'the father of Chemistry' is credited with discovering sulphuric acid about 1200 years ago.

Key points

- Sulphur is burnt to make sulphur dioxide, which is used in the production of sulphuric acid.
- Other uses of sulphur are vulcanisation of rubber, making polymers and in agriculture.
- Sulphur dioxide is used as a food preservative, in sulphuric acid manufacture and as a bleach.
- The main use of sulphuric acid is in making fertilisers.
- Sulphuric acid is used to make detergents and dyes and for cleaning metal surfaces.
- The impact of the sulphuric acid industry on the environment is largely the formation of acid rain.

Revision questions

Answers to all revision questions can be found on the accompanying CD.

1 List four factors that determine the location of an industrial chemical plant.

2 a Give the name and formulae of two bauxite ores from which aluminium is extracted.
   b i What is the name and formula of the impurity that gives some bauxite its red colour?
      ii Describe how the impurity in i above is removed from the bauxite.
   c By writing balanced equations, show how silicon dioxide is removed from the bauxite.
   d Why is cryolite added to aluminium oxide during its electrolysis?
   e Write the equations for the two reactions that occur at the cathode and anode during the electrolysis of aluminium oxide.
   f Why must the anode be replaced periodically?
   g Why is the electrolytic process not carried out by the bauxite plants in the Caribbean?

3 a List three effects that the aluminium industry has on the environment.
   b What are the properties (physical or chemical) that allow aluminium to be used to make the following:
      i overhead electricity cables
      ii food packaging
      iii cooking pots
      iv fire fighter clothing

4 a i What is the name of the process by which crude oil is separated into its components called?
      ii Describe the principles upon which the separating method mentioned in a i above is based.
   b List four fractions of crude oil and state the uses of each.

5 a i Define 'cracking' and state the two types.
      ii Why do petroleum companies carry out the process of cracking?
   b An alkane consisting of 12 carbon atoms undergoes cracking to produce a saturated hydrocarbon with 10 carbon atoms and an unsaturated hydrocarbon.
      i Write an equation to show this reaction.
      ii Name the unsaturated hydrocarbon formed.

6 a What is reforming?
   b Why is reforming a useful process in the petroleum industry?

7 List three effects that the petroleum industry has on the environment.

8 a Write a balanced equation for the Haber Process showing the production of ammonia.
   b i How is the hydrogen required for the process obtained from methane? Include a balanced equation in your answer.
      ii How is the nitrogen required for the process obtained?
   c i State the conditions under which ammonia is produced industrially using the Haber Process.
      ii Using Le Chatelier's principle state the conditions of pressure and temperature that would produce a maximum yield of ammonia using this process.
      iii Explain your answer in c ii above.
      iv Why are the conditions used for the process, as stated in c i not exactly in line with the conditions dictated by Le Chatelier's principle, as stated in c ii?

9 a List four uses of ammonia.
   b Explain the process of eutrophication and the role of ammonium-based fertilisers in contributing to this process.

10 a Write a balanced equation showing the fermentation of glucose to produce ethanol.
    b What is name of the enzyme used in the fermentation process in a above?
    c Theoretically, higher temperatures would increase the rate of production of ethanol using this process, however lower temperatures are chosen. Give a reason for this.
    d State two other conditions (besides the temperature and the presence of the enzyme) that are required for the fermentation process.
    e Name the process by which the concentration of the ethanol produced by fermentation is increased.
    f List three uses of ethanol.

11 a Why is ethanol classified as a drug?
    b State three short-term effects of ethanol on the body.
    c State two long-term effects of ethanol on the body.
    d State one social and one economic consequence of ethanol.

12 Write the equations for the reactions taking place at the anode and cathode in the diaphragm cell for the production of chlorine from brine.

13 a Why is the industry called the chloro-alkali industry?

b State **four** uses of chlorine or its products.

c Why does the asbestos used in the diaphragm have to be changed on a regular basis?

d Under the following headings, discuss the impact of the production of chlorine on the environment:

   i Sodium hydroxide

   ii The ozone layer

   iii Polyvinyl chloride

   iv Dioxins

14 State the **three** stages used in the industrial preparation of sulphuric acid. Write balanced equations for each stage.

15 a Using Le Chatelier's principle state the conditions of pressure and temperature that would produce a maximum yield of sulphur trioxide.

b Explain your answer in **a** above.

c State the conditions used for the industrial production of sulphur trioxide.

16 a List **two** uses of sulphur dioxide.

b List **four** uses of sulphuric acid.

c State **two** effects of the sulphuric acid industry on the environment.

14 Chemistry and the environment

14.1 The water cycle and water purification

Learning outcomes

On completion of this section, you should be able to:

- describe the importance of the water cycle
- describe methods for purifying water (desalination, fractional distillation, electrodialysis)
- explain the importance of dissolved oxygen to aquatic life.

The water cycle

On our planet, water only stays in the same place when frozen in permanent ice caps or when stored in aquifers (huge natural reservoirs of water absorbed into porous rocks). The water on the surface of the Earth and in the atmosphere is constantly evaporating and condensing to form a **water cycle** (Figure 14.1.1).

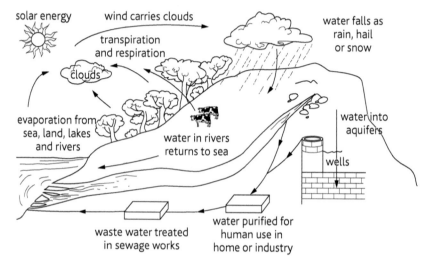

Fig 14.1.1 *The water cycle*

- Fresh water evaporates from seas, lakes and the soil.
- The water vapour is carried by winds across the Earth's surface.
- When this air reaches colder land masses or colder air, the water vapour cools and condenses to tiny droplets of water to form clouds.
- The droplets get larger and fall as rain or snow. Some of the water falls back into the sea or lakes and some falls on the land.
- The water falling onto the land drains through the soil into streams and rivers and finds its way back into the sea to start the cycle again.
- In this way, the amount of water in the seas, the atmosphere and land is kept constant.

Dissolved oxygen and aquatic life

Aquatic life such as fish, crabs and plankton (tiny animals in the sea) need dissolved oxygen (DO) for respiration. If there is not sufficient oxygen dissolved in the water, aquatic animals will die. Oxygen gets into rivers, lakes and the sea by:

- diffusion through the water surface from the air
- diffusion from bubbles of air trapped in fast-flowing water as it goes over waterfalls or from photosynthesis of aquatic plants.

At 20 °C and at atmospheric pressure there is about 9.5 mg dm^{-3} DO. The amount of DO in water depends on several factors:

- The DO decreases as temperature increases and as pressure decreases.

Did you know?

If the total dissolved gases in water is too high fish can get 'gas bubble disease'. This is rather like the 'bends' which affects deep sea divers. Bubbles of gas form in the blood and this can block the flow of blood through the blood vessels. It can eventually be fatal.

- Salt water has less DO than freshwater.
- Degree of agitation of the water surface. Stagnant water has less DO than flowing water.
- The number of bacteria and water plants removing oxygen from the water e.g. in eutrophication (see Section 13.2).

The amount of DO needed by an organism depends on the species. For example fish such as trout and salmon need relatively higher amounts to survive than small invertebrates. The minimum DO level to support a healthy fish population is $4-5\,mg\,dm^{-3}$. When the DO levels fall below this, aquatic life is put under stress. At DO below $1-2\,mg\,dm^{-3}$ fish will die.

Purifying water

Pure water for drinking or industry can be obtained from seawater or brackish water (slightly salty water) by **desalination**. Desalination is the removal of salts from the water, it is especially important in countries where fresh water is in short supply, e.g. those having low rainfall.

Fractional distillation and flash distillation

Pure water can be obtained from seawater by fractional distillation but this is expensive as it requires a lot of energy. This is most suited to countries where energy supplies are cheap. Flash distillation produces over 60% of all the desalinated water in the world. Seawater is fed in at one end of a large tank and warmed. It is then pumped through a series of compartments (heat exchangers) where the pressure and temperature is successively lower. At each of these stages, some of the seawater turns to steam and condenses again, so forming pure water (see Figure 14.1.2).

Electrodialysis

Electrodialysis is used to transport the ions in salt (Na^+ and Cl^-) from one solution to another through an ion-exchange membrane under the influence of a voltage. The Cl^- ions move towards the anode. These ions pass through a positively charged anion-exchange membrane but are prevented from further migration to the anode by the negatively charged cation-exchange membrane. The Na^+ ions move to the cathode but are prevented migrating further by an anion-exchange membrane. So the ions are concentrated in one part of the cell leaving the other part depleted in ions.

Ion exchange

Ion-exchange columns have been used in commercial and household water purification units for many years. These are based on the idea that an ion on the resin swaps with an ion in solution. In order to remove sodium and chloride ions from water a series of three columns is used, one which acts as the desalination column and another two which serve to maintain a specific balance of anions and cations.

Reverse osmosis

Water is forced from a region of high solute concentration (salt solution) to a region of low solute concentration through a selectively permeable membrane, by applying a pressure on the high concentration side. The salt solution gets more concentrated as the water passes from it.

Freeze desalination

When salty water freezes, the ice separates leaving a solution with a high concentration of salts. The ice is taken from the solution and re-melted. By repeating this process several times, ice free of salts is formed.

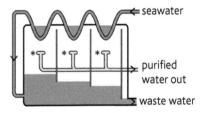

Fig 14.1.2 *Simplified diagram of a flash distillation unit (* shows where the water condenses)*

Key points

- The amount of water in the atmosphere, the seas and the land is kept constant by the water cycle.

- Dissolved oxygen is essential for aquatic life.

- Temperature, salt and number of bacteria all affect the amount of oxygen dissolved in water.

- Water can be desalinated by distillation, electrodialysis, ion exchange, **reverse osmosis** and freeze desalination.

The sources of water pollution

Water pollution is due to the introduction of chemical or biological materials into the water. Industrial wastes are often discharged into seas and rivers and, although often treated, still have toxic substances in them. Domestic sewage also contains synthetic detergents which may end up in the sea. Material may also leach into rivers from waste disposal sites. Common pollutants include:

Nitrates: from the leaching of fertilisers (see Section 13.2).

Phosphates: from the leaching of fertilisers and sewage disposal including phosphates from detergents.

Heavy metals: e.g. mercury from the chlor-alkali industry, lead from old water pipes and from anti-knock agents from petrol, cadmium from electroplating and batteries.

Cyanides: from metal extraction industries, e.g. silver, gold, the iron and steel industry and from the discharge of material from the preparation of organic chemicals.

Other metals: e.g. aluminium from clay extraction, trace elements by leaching through the soil or from discharges from industry.

Pesticides and herbicides: rain washes these from crops and leaches them into rivers.

Petroleum residues: (See Section 12.2).

Suspended particles: Small particles of clay and other material such as paints washed from construction sites, quarries and storm sewers.

Pollutants and the aquatic environment

- Suspended solids in the water can reduce the amount of sunlight reaching water plants. This may inhibit photosynthesis and even lead to plant death. This disturbs the food chains in the water.
- Eutrophication through fertiliser run-off (see Section 13.2) can lead to the death of organisms. Phosphates from sewage and detergents also cause eutrophication.
- Toxicity of inorganic waste. Many heavy metals are poisonous, e.g. cadmium, mercury and lead. Their concentrations can reach high levels in enclosed bodies of water.
- Microbial contamination of water is responsible for 80% of all the sickness in the world. Diseases such as cholera, typhoid and malaria are associated with contaminated water.
- Many pesticides and herbicides are halogenated hydrocarbons. Their presence in water can result in the death of invertebrates (and its consequent effect on the food chain) as well as infertility in birds.
- Oil spillages. If they leak into water, petroleum residues may have a harmful effect on wildlife, e.g. seals can be blinded and birds may die (see Section 12.2).

Testing for selected pollutants

Nitrates

Add aqueous sodium hydroxide to the suspected nitrate and then either zinc powder or aluminium powder (or Devarda's alloy). On warming, ammonia gas is released (see *Unit 1 Study Guide*, Section 14.3).

Phosphates

Acidify with concentrated nitric acid and add a little ammonium molybdate. The formation of a bright yellow precipitate on warming gently indicates that a phosphate is present.

Lead ions

Add $1\,mol\,dm^{-3}$ hydrochloric acid. The presence of a white precipitate that redissolves in hot water indicates that lead is present. A confirmatory test is to add aqueous potassium iodide to the acidified water. A bright yellow precipitate indicates the presence of lead.

Cyanide

Add iron(II) sulphate to the solution then acidify with hydrochloric acid. If a deep blue complex ion is formed, cyanide ions are present.

Turbidity

Turbidity refers to the cloudiness of suspended matter in a liquid. The simplest way to measure turbidity is to measure the light transmitted through a column containing the liquid under test. A better method is to use a nephelometer to measure the light scattered by the suspended particles using a detector on the same side of the tube as the light beam. The greater the number of suspended particles, the greater is the scattering and the greater the detector reading.

Water treatment

Water has to be treated to make it safe for drinking. The main steps in this process to purify dirty water are:

- *Screening:* Removes large floating objects.
- *Aeration:* Removes volatile substances such as hydrogen sulphide and volatile oils.
- *Flocculation and sedimentation:* The water is agitated. Small particles clump together and are then removed after settling.
- *Filtration:* Removes finely suspended particles from the water.
- *Coagulation:* Iron sulphate or aluminium sulphate are added to help very fine suspended particles clump together.
- *Disinfection:* Chlorine is added to kill bacteria and other microorganisms.
- *Adsorption:* Activated charcoal is used to adsorb organic chemicals which might give a bad odour and taste to the water.
- *Oxidation:* Undesirable substances, e.g. cyanide-containing compounds, are oxidised with ozone to form less harmful products.
- *Desalination:* (see Section 14.1).

Did you know?

Mangroves are areas where various kinds of trees grow in several feet of water along the coast in some tropical and subtropical areas. In mangrove swamps high turbidity is needed in the water to protect some young fish from predators and preserve the ecosystem.

Key points

- Common pollutants in water include nitrates, phosphates, heavy metals, cyanide, pesticides, petroleum residues and suspended particles.
- There are specific chemical tests for nitrates, phosphates, lead ions and cyanide.
- Turbidity is measured by the ability of a suspension to scatter light.
- Water is treated by using the processes of aeration, filtration, coagulation, disinfection and charcoal adsorption.

14.3 Ozone in the atmosphere

Learning outcomes

On completion of this section, you should be able to:

- explain how the concentration of ozone in the atmosphere is maintained

- understand the term 'photodissociation'

- describe the environmental significance of CFCs in the ozone layer

- describe some free radical reactions in the upper atmosphere

- describe the effects of ozone on human life (referring to the stratosphere and troposphere).

Ozone in the atmosphere

The stratosphere is the part of the atmosphere about 20–50 km above the Earth. Ozone, O_3, is present in the stratosphere in a 'layer' which varies in thickness. The ozone which is present at a concentration of about 10 parts per million, absorbs harmful ultraviolet (UV) radiation from the Sun.

The importance of the ozone layer to human health

The **ozone layer** is important for human health because if too much UV light reaches surface of Earth:

- there is an increased risk of sunburn and skin cancer
- the skin ages faster
- people are more likely to get cataracts in their eyes
- we may have reduced resistance to some diseases.

Ozone formation in the stratosphere

In the stratosphere, ozone is formed naturally by the action of UV light on oxygen. This is a **photodissociation** reaction – a reaction in which light (usually ultraviolet light) causes bond breaking. The UV light has enough energy to cause oxygen molecules to dissociate to form oxygen free radicals, which are very reactive. This is an example of homolytic fission (see Section 2.1).

$$O_2(g) \xrightarrow{\text{UV light}} 2O\bullet(g)$$
$$\text{oxygen radical}$$

An oxygen radical can react with an oxygen molecule to form ozone.

$$O_2(g) + O\bullet(g) \to O_3(g)$$

Ozone is also broken down by UV light from the Sun. Oxygen and an oxygen free radical are formed.

$$O_3(g) \xrightarrow{\text{UV light}} O_2(g) + O\bullet(g)$$

In the absence of any other factors decomposing the ozone, the rate of formation of ozone is in balance with the rate of breakdown. So the amount of ozone in the atmosphere remains constant.

Ozone-depleting chemicals

Chlorofluorocarbons (CFCs) used as refrigerants and aerosols are not toxic and are unreactive under normal conditions. They persist in the atmosphere for hundreds of years. After many years, CFC molecules may reach the stratosphere, where the UV light can decompose these molecules. Highly reactive free radicals are formed by homolytic fission. A cycle involving **initiation**, **propagation** and **termination** occurs (see Section 2.1). We shall use the chlorofluorocarbon, CCl_2F_2 as an example.

Initiation: The UV light is strong enough to break the C—Cl bond but not the C—F bond.

$$CCl_2F_2 \xrightarrow{\text{UV light}} Cl\bullet + \bullet CClF_2$$

Did you know?

In 1785 Dutch scientist Martin van Marum was the first person to record a distinctive smell when electrical machinery was working. This was later found to be due to ozone. The word ozone comes from the Greek 'ozein' which means to smell.

Propagation: Cl free radicals can then attack ozone molecules, e.g.

$$Cl\bullet + O_3 \rightarrow ClO\bullet + O_2$$
$$ClO\bullet + O_3 \rightarrow Cl\bullet + 2O_2$$

The result of these reactions is that ozone is converted to oxygen.

$$2O_3(g) \rightarrow 3O_2(g)$$

The Cl• radical acts as a catalyst because it is constantly being reformed. In these chain reactions, a chlorine radical may break down about 100 000 ozone molecules before a termination reaction between two free radicals occurs. The presence of CFCs in the atmosphere therefore leads to **ozone depletion**. Other chloro-organic compounds may also have this effect, e.g. tetrachloromethane and trichloroethane. Nitrogen oxides from high flying jet aircraft may also contribute to ozone depletion.

The effects of low level ozone

Ozone present in the troposphere (the layer of the atmosphere next to Earth's surface) can be toxic to plant and animal life. It can:

- irritate the respiratory system and cause breathing difficulties. It has been linked to increased incidences of asthma and bronchitis.
- have a bad affect on the heart and blood vessels. It may cause cholesterol-like compounds to form in the lungs. These may cause cardiovascular problems such as atherosclerosis (hardening of the arteries).

Ozone is one of the factors contributing to photochemical smog. Nitrogen dioxide from car exhausts (see Section 14.8) can undergo photolytic reactions in the presence of UV light. Ozone is formed during this cycle and nitrogen dioxide is regenerated.

$$NO_2 \xrightarrow{\quad UV\ light \quad} NO\ +\ \quad O\bullet$$

nitrogen   nitric   oxygen free
dioxide   oxide   radical

$$O\bullet + O_2 \rightarrow O_3$$
ozone

$$NO + O_3 \rightarrow NO_2 + O_2$$

When hydrocarbons and/or carbon monoxide are present, this cycle gets disrupted and ozone reacts with unsaturated hydrocarbons to produce organic radicals. These are responsible for many of the harmful effects of photochemical smog (see Section 14.8).

Key points

- The concentration of ozone in the atmosphere is maintained by a cycle involving free radical reactions.
- Photodissociation is the breaking of a bond by light, usually UV light.
- CFCs can deplete the ozone layer by catalytic reactions involving free radicals and initiated by ultraviolet light.
- Stratospheric ozone protects humans from the harmful UV rays of the Sun.
- Tropospheric ozone can be produced by the photochemical decomposition of nitrogen oxides and subsequent reaction of an oxygen free radical with oxygen.
- Ozone is harmful to plant and animal life.

✓ Exam tips

The effect of CFCs on the depletion of the ozone layer and the production of ozone in the troposphere as a result of the interaction of UV light with nitrogen oxides from car exhausts are both examples of chain reactions. Make sure that you know the initiation, propagation and termination steps related to these reactions. We first came across these in Section 2.1.

14.4 The carbon cycle

Carbon sinks

The oceans, the atmosphere and rocks all contain carbon. We call these carbon sources *carbon reservoirs* or *carbon sinks*. The amount of carbon in each of these reservoirs has not changed much over millions of years. This is because there is a balance between the uptake and release of carbon from these reservoirs. The reservoir of carbon in the atmosphere, mainly as carbon dioxide, is the one which can undergo most rapid changes.

Releasing carbon into the atmosphere

There are several ways by which carbon gets into the atmosphere.

Respiration: This is the process in which food is oxidised in the body by a complex series of reactions to produce energy. During this process carbon dioxide is released into the atmosphere. Aerobic respiration removes oxygen from the atmosphere to do this.

$$C_6H_{12}O_6(aq) + 6O_2(g) \rightarrow 6CO_2(g) + 6H_2O(l)$$

Combustion of fuels: Fuels such as coal, wood and hydrocarbons produce CO_2 when they burn. This CO_2 escapes into the atmosphere. When this happens, oxygen is also removed from the atmosphere. Combustion can be due to human activity (e.g. for transport) or natural activity (e.g. forest fires).

Other decompositions: Small amounts of carbon are released into the atmosphere by the breakdown of vegetation in swamps.

From oceans: When CO_2 dissolves in seawater various equilibria are set up involving hydrogencarbonate and carbonate ions, e.g.

$$\underset{\text{hydrogencarbonate}}{2HCO_3^-(aq)} \rightleftharpoons CO_2(g) + H_2O(l) + \underset{\text{carbonate}}{CO_3^{2-}(aq)}$$

Most gases are less soluble in water as the temperature increases. When water in the oceans gets warmer CO_2 bubbles out of solution and the equilibrium above is shifted to the right.

The uptake of carbon from the atmosphere

There are two main ways in which carbon dioxide is removed from the atmosphere: photosynthesis and by the oceans.

Photosynthesis: Plants remove carbon dioxide from the atmosphere during photosynthesis to make carbohydrates.

$$\underset{\text{glucose}}{6CO_2(g) + 6H_2O(l) \rightarrow C_6H_{12}O_6(aq) + 6O_2(g)}$$

The energy for this reaction comes from sunlight. Chlorophyll, the green pigment in plants traps the sunlight and acts as a catalyst.

By oceans: CO_2 is quite soluble in water and large amounts are removed from the atmosphere when it dissolves in oceans:

$$CO_2(g) + H_2O(l) \rightleftharpoons HCO_3^-(aq) + H^+(aq)$$

Balancing uptake and release of CO_2

The complete carbon cycle is shown in Figure 14.4.1.

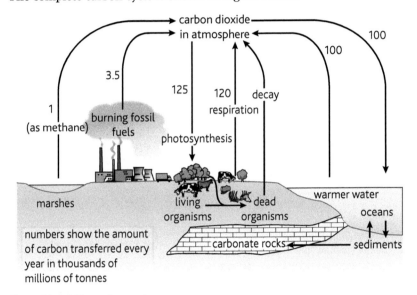

numbers show the amount
of carbon transferred every
year in thousands of
millions of tonnes

***Figure 14.4.1** The carbon cycle*

The two processes which keep this cycle in balance are respiration and photosynthesis. Respiration releases CO_2 into the air and takes up oxygen. Photosynthesis takes up CO_2 from the air and puts oxygen back into the air. These two processes are approximately balanced so that the CO_2 content of the air remains fairly constant. In addition, the large amount of CO_2 taken up by the oceans is balanced by that released from the oceans.

Upsetting the balance

The burning of fossils fuels only releases a small amount of carbon dioxide into the atmosphere compared with the amount released by respiration. If fossil fuels were being formed as fast as they were being burnt there would not be a problem. But fossil fuels are not being formed. So we are putting more CO_2 into the atmosphere.

Many people are worried that an increase in the amount of CO_2 in the atmosphere will put the carbon cycle out of balance and increase global warming (see Section 14.5).

As the world's population increases, many forests are being cleared because of an increased need of land for agriculture, quarrying or housing. This deforestation means that less CO_2 is being removed from the atmosphere by photosynthesis. So the balance between photosynthesis and respiration is altered. That is why **reforestation** (replanting of trees to replace those lost) is important in maintaining this balance.

Exam tips

The two most important regulating features of the carbon cycle are respiration and photosynthesis. Anything that drastically reduces or increases the extent to which one or other of these takes place will upset the carbon cycle.

Did you know?

HCO_3^- ions and CO_3^{2-} ions can be taken up by tiny animals (plankton) in the sea. After they die, plankton fall to the sea bed. Over millions of years these dead organisms form carbonate rocks, e.g. limestone. This carbon is taken out of the carbon cycle unless the carbonate rocks are heated to make lime.

Key points

- The carbon cycle keeps the level of carbon dioxide in the atmosphere relatively constant.

- The carbon dioxide released into the atmosphere by the respiration in living organisms is balanced by the uptake of carbon dioxide in photosynthesis.

- Burning fossil fuels and deforestation may cause the carbon cycle to become unbalanced.

- Reforestation helps to rebalance the carbon cycle.

On completion of this section, you should be able to:

- explain the terms 'greenhouse effect' and 'global warming'
- describe the re-radiation of energy from the infrared region into the atmosphere.

The greenhouse effect

The **greenhouse effect** is a process by which thermal radiation from the Sun is absorbed by the atmosphere and re-radiated in all directions so that the temperature of the Earth's surface and atmosphere is higher than if it were heated directly by radiation from the Sun.

A simplified diagram of the greenhouse effect is shown in Figure 14.5.1.

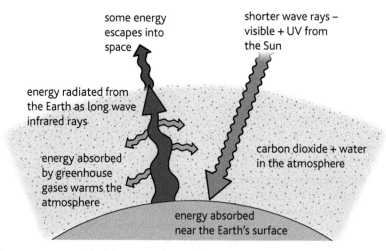

Figure 14.5.1 *Simplified diagram of the greenhouse effect*

In the greenhouse effect, gases in the atmosphere such as carbon dioxide and water vapour prevent the Earth from cooling down too rapidly when it is not exposed to the Sun's rays. It works like this:

- Ultraviolet and visible radiation from the Sun have relatively short wavelengths. The visible radiation and some of the UV radiation pass through the atmosphere without being absorbed by carbon dioxide.
- The short wavelength radiation hits the Earth's surface so that the Earth's surface gains energy.
- When the Earth's surface absorbs the short wavelength rays it heats up.
- Energy is lost from the surface as radiation with a longer wavelength. The radiation emitted is in the infrared region.
- Infrared radiation can be absorbed by greenhouses gases such as CO_2 and water, which are naturally present in the atmosphere.
- Some of the heat is re-radiated back to Earth and the lower layers of the atmosphere and some escapes into space. Less radiation escapes into space than would be the case if greenhouse gases such as carbon dioxide were not present.
- The absorbed radiation raises the temperature of the atmosphere. This natural raising of the atmospheric temperature is called global warming.

Global warming

Global warming is the rise in the temperature of the Earth's atmosphere that arises from the greenhouse effect. If we did not have global warming, the Earth would be extremely cold, because the radiation would be reflected away from the surface rather than being absorbed by the atmosphere. The more carbon dioxide and water vapour there is in the atmosphere, the more heat is absorbed and re-radiated back to the Earth and to the atmosphere because of the greenhouse effect. So the atmosphere heats up more. There is enhanced global warming (an increase in global warming).

Greenhouse gases

A **greenhouse gas** absorbs and emits radiation in the infrared part of the electromagnetic spectrum. The absorption spectrum of carbon dioxide, a naturally-occurring greenhouse gas, is shown in Figure 14.5.2.

The main greenhouse gases naturally present in the atmosphere are:

- *Water vapour:* This may contribute from about 30–60% of the greenhouse effect especially in the lower layers of the atmosphere. It is present in the atmosphere in small amounts, which are kept relatively constant by the water cycle.

- *Carbon dioxide:* CO_2 is naturally present in the atmosphere. Although it is only present in low concentrations, it is a potent greenhouse gas contributing to at least 30% of the greenhouse effect.

- *Methane:* This is found in the atmosphere at much lower concentrations than CO_2, but it absorbs relatively more IR radiation and may contribute as much as 5–10% of the greenhouse effect. Methane is formed by the action of bacteria in the digestive system of animals and by the bacterial decay in marshes as well as from rice paddy fields.

Ozone, CFCs and nitrogen oxides are also effective greenhouse gases.

Enhanced global warming and climate change

Over the past 150 years the amount of CO_2 in the atmosphere has been increasing because of the increased burning of fossil fuels in power stations and for transport. The present percentage of CO_2 in the atmosphere is about 0.039%, whereas 50 years ago it was about 0.031%. The concentrations of other greenhouse gases, such as methane and nitrogen oxides and tropospheric ozone, have been increasing in the atmosphere in recent years for the same reasons. The increased concentration of greenhouse gases leads to increased global warming. A warmer atmosphere can affect our climate by:

- melting polar ice caps and hence raising sea levels leading to increased flooding of low-lying areas

- reducing the rainfall in some areas so increasing the rate of formation of deserts

- making the weather more violent and unpredictable

- increasing the temperature of the oceans leading to the death of some species such as corals. An increased amount of CO_2 will be also be released from the oceans leading to an even further increase in global warming.

✓ *Exam tips*

You must be able to distinguish between the terms *greenhouse effect* and *global warming*. The greenhouse effect relates to the way radiation is absorbed by the Earth's surface and the atmosphere close to the Earth's surface, then re-emitted as longer range radiation. Global warming refers to the temperature increase arising from the greenhouse effect.

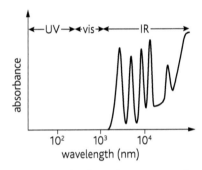

Fig. 14.5.2 *The absorption spectrum of carbon dioxide*

Key points

- The greenhouse effect is a process by which thermal radiation is absorbed by the atmosphere and re-radiated so that the temperature of the Earth's surface and atmosphere is higher than if it were heated directly by radiation from the Sun.

- Global warming is the rise in the temperature of the Earth's atmosphere which arises from the greenhouse effect.

- A greenhouse gas absorbs and emits radiation in the infrared part of the electromagnetic spectrum.

Learning outcomes

On completion of this section, you should be able to:

- describe the nitrogen cycle
- explain how the atmospheric concentrations of the oxides of nitrogen may be altered.

The nitrogen cycle

Nitrogen gas forms 78% of the atmosphere by volume, but it has low chemical reactivity. Nitrogen can be recycled at a sufficient rate in the atmosphere to allow plants to grow. Figure 14.6.1 shows a simplified diagram of the **nitrogen cycle**.

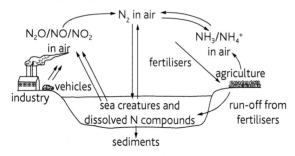

Figure 14.6.1 *Simplified diagram of the nitrogen cycle*

Much of the nitrogen cycle is dominated by reactions involving enzyme-catalysed conversions in microbes, plants and animals. These reactions are shown in Figure 14.6.2. Many of the reverse reactions can be catalysed by microorganisms.

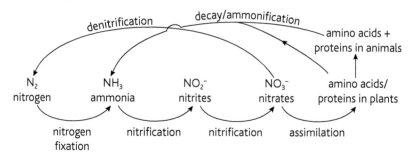

Figure 14.6.2 *Some biological oxidations and reductions involving nitrogen compounds*

✓ Exam tips

The nitrogen cycle is very complex. You need only learn the basic reactions below rather than all the biological conversions of ammonia to proteins.

Reactions which release nitrogen into the atmosphere

- **Denitrification:** Denitrifying bacteria occur under both aerobic and anaerobic conditions in the soil and in the oceans. They use organic compounds to reduce nitrates to nitrogen gas. The organic compounds are oxidised to carbon dioxide.

$$5CH_2O(aq) + 4NO_3^-(aq) + 4H^+(aq) \rightarrow 2N_2(g) + 5CO_2(g) + 7H_2O(l)$$

(where CH_2O is a simplified formula for a sugar)

The nitrates arise from fertilisers, decomposition of animal and plant remains or from conversion of nitrogen oxides into nitrates.

- *Ammonium oxidation:* Nitrogen gas can be reformed by anaerobic oxidation of ammonia and ammonium ions by bacteria. This occurs largely in the sea. The ammonium ions arise from decaying plant and animal remains and fertiliser run off.

Reactions which remove nitrogen from the atmosphere

- *The* **Haber Process**: This removes nitrogen directly from the atmosphere for ammonia synthesis. The nitrogen is then used to make fertilisers and nitric acid.

- *Lightning:* The high temperatures of lightning cause oxygen and nitrogen to combine. Nitrogen oxides are formed, which can undergo further reactions in the atmosphere, being finally removed as nitrates which dissolve in the sea or in groundwater.

- **Nitrogen fixation:** Certain bacteria or blue-green algae can convert atmospheric nitrogen into ammonia. The nitrogen-fixing bacteria can be aerobic or anaerobic. The ammonia can then be converted to nitrates by nitrifying bacteria in the soil. Plants then absorb the dissolved nitrates to synthesise amino acids and proteins (see Figure 14.6.2).

- *Car engines:* The high temperature and pressure inside a car engine can cause nitrogen and oxygen to combine. A mixture of nitrogen oxides is formed, which are emitted from the exhaust gases. This mixture is sometimes called NO_x to show that several different gases are formed:

$$N_2(g) + O_2(g) \rightarrow 2NO(g)$$

$$N_2(g) + 2O_2(g) \rightarrow 2NO_2(g)$$

- *High temperature furnaces:* The high temperature in these furnaces causes nitrogen and oxygen to combine to form nitrogen oxides.

The balance of the nitrogen cycle

The atmosphere has a very large sink of nitrogen, so its concentration remains more or less constant. Before the days of the Haber process, the main parts of the nitrogen cycle were the fixation of nitrogen by microorganisms, which was roughly balanced by the natural processes that put nitrogen back into the atmosphere. The removal of nitrogen from the atmosphere for the production of fertilisers is now about the same as the amount removed by bacterial nitrogen fixation. More nitrogen is also being removed from the air due to the formation of nitrogen oxides from vehicle exhausts and industry. Although these processes at the present time may not affect the working of the nitrogen cycle, the presence of increasing amounts of nitrogen oxides being put into the atmosphere by vehicles and by industry poses particular problems (see Sections 14.7 and 14.8).

Nitrogen oxides in the atmosphere

Oxides of nitrogen are recycled in the atmosphere by natural processes. Nitrous oxide, N_2O is added to the atmosphere by denitrification reactions. Nitric oxide, NO, and nitrogen dioxide, NO_2, are produced during thunderstorms:

$$N_2(g) + O_2(g) \rightarrow 2NO(g) \text{ and } N_2(g) + 2O_2(g) \rightarrow 2NO_2(g)$$

Nitrous oxide passes from the lower atmosphere to the stratosphere, where it is converted to N_2 and NO by photolytic reactions initiated by UV light:

$$N_2O(g) \rightarrow NO(g) + N\bullet(g)$$

$$N_2O(g) \rightarrow N_2(g) + O\bullet(g)$$

Nitric oxide can form by the reaction of N_2O with the oxygen free radicals formed by the latter reaction or from the decomposition of ozone:

$$N_2O(g) + O\bullet(g) \rightarrow 2NO(g)$$

The NO formed can catalyse the decomposition of ozone. Some NO can also react with either oxygen free radicals or ozone to form acidic NO_2 which can contribute to acid rain (see Section 14.7).

This part of the nitrogen cycle is of concern to many scientists because humans are affecting it to a considerable extent (see Sections 14.7 and 14.8).

Did you know?

Some nitrogen-fixing bacteria are found in swellings on the plant roots of beans and clover plants. These swellings are called root nodules. Farmers sometime plough crops of beans or clover back into the soil because they increase the nitrogen content of the soil.

Key points

- Processes that remove nitrogen from the atmosphere include the Haber process, lightning and nitrogen fixation.

- Processes that add nitrogen to the atmosphere include denitrification and ammonia oxidation.

- The concentration of nitrogen oxides in the atmosphere is increasing because of nitrogen oxides produced by vehicle emissions and high temperature furnaces.

On completion of this section, you should be able to:

- describe the effects of the products of combustion of fuels containing sulphur
- explain how acid rain is formed from oxides of sulphur and nitrogen.

What is acid rain?

Rain is naturally slightly acidic due to carbon dioxide reacting with water in the air. This rain has a pH of about 5.6. But this is not classed as acid rain. If the acidity of the rain falls below about pH 5, the rain is called **acid rain**. This is caused by oxides of sulphur and nitrogen in the atmosphere reacting with water vapour.

Acidic oxides in the air

Coal and natural gas contain some sulphur impurities. Fuels for transport also contain small amounts of sulphur, although most of it is removed before the fuel is sold. When these fuels are burnt, the sulphur is oxidised to sulphur dioxide, which is an acidic gas:

$$S(s) + O_2(g) \rightarrow SO_2(g)$$

Volcanoes are a natural source of sulphur dioxide. They produce nearly a third of the sulphur dioxide which pollutes the atmosphere.

Nitrogen oxides can also get into the air from car exhausts and from high temperature furnaces. Nitrous oxide and nitric oxide are not acidic but nitrogen dioxide is an acidic gas.

The formation of acid rain

The formation of acid rain involves two stages: oxidation and deposition. Figure 14.7.1 shows these stages.

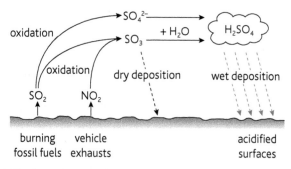

Figure 14.7.1 *The formation of acid rain*

Oxidation reactions in the atmosphere

Sulphur dioxide in the atmosphere is oxidised by a variety of catalysts.

Oxidation by nitrogen dioxide: This can take place fairly quickly in the atmosphere. Sulphur dioxide reacts with nitrogen dioxide to form sulphur trioxide and nitric oxide. The nitrogen dioxide is then reformed by reaction of nitric oxide with oxygen in the air.

$$SO_2(g) + NO_2(g) \rightarrow \underset{\text{sulphur trioxide}}{SO_3(g)} + \underset{\text{nitric oxide}}{NO(g)}$$

then: $\quad 2NO(g) + O_2(g) \rightarrow 2NO_2(g)$

The NO_2 which is reformed can go on to oxidise another SO_2 molecule. This process can be repeated to oxidise many more SO_2 molecules. The NO_2 acts as a catalyst.

Oxidation involving free radials: These reactions often take place higher in the atmosphere and so do not occur until the SO_2 has moved up to these levels. Examples are:

- Oxidation with oxygen free radicals or ozone:

$$SO_2(g) + O_3(g) \rightarrow SO_3(g) + O_2(g)$$

- Oxidation to sulphates by OH• radicals. The OH• radicals are formed by the action of O• radicals from ozone with water $(H_2O + O• \rightarrow 2OH•)$.

$$OH•(g) + SO_2(g) \rightarrow HSO_3•(g) \xrightarrow{\text{several steps}} H_2SO_4$$

Nitric oxide from car exhausts can also react with either oxygen, free radicals or ozone to form the acidic gas, nitrogen dioxide, NO_2.

$$NO(g) + O•(g) \rightarrow NO_2(g) \text{ or } NO(g) + O_3(g) \rightarrow NO_2(g) + O_2(g)$$

Acid formation (deposition)

Wet deposition: Sulphur trioxide and small particles of sulphates in the atmosphere react with or dissolve in water vapour in the atmosphere to form a dilute solution of sulphuric acid. This falls with the rain to form acid rain.

$$SO_3(g) + H_2O(l) \rightarrow H_2SO_4(aq)$$

Nitrogen dioxide in the atmosphere reacts with water vapour in the atmosphere to from a solution of nitric acid.

$$2NO_2(g) + H_2O(l) \rightarrow HNO_3(aq) + HNO_2(aq)$$

Some sulphur dioxide may also react directly with water to form sulphurous acid when it is raining.

$$SO_2(g) + H_2O(l) \rightarrow H_2SO_3(aq)$$

Dry deposition: Small particles of sulphates and nitrates can form in the air when nitric or sulphuric acid reacts with ammonia in the atmosphere. These compounds and dry SO_2 and SO_3 gases can be deposited on moist surfaces such as plants and wet buildings to form acids.

The effects of acid rain

- Trees may have their leaves and roots damaged. This can lead to forest death.
- Lakes and rivers become acidic. Some aquatic organisms may die.
- Soil may become too acidic to grow crop plants and minerals may be leached out of the soil.
- Buildings made from carbonate rocks may be eroded.
- Metals structures such as bridges and railings may corrode.

Acid rain may fall far from the source of pollution. Sulphur dioxide, sulphur trioxide and nitrogen oxides can be carried by winds as far as 200 km from their sources. So the effects on the environment will not always be obvious close to the sources.

Key points

- When fuels containing sulphur are burnt, sulphur dioxide is formed.
- Sulphur dioxide is oxidised in the atmosphere by free radicals or nitrogen dioxide to form sulphur trioxide and sulphates.
- Sulphur trioxide reacts with water vapour to form dilute sulphuric acid in the rain.
- Nitrogen oxides can be oxidised to form nitrogen dioxide, which dissolves in water vapour to form nitric acid.

Learning outcomes

On completion of this section, you should be able to:

- describe the difference between primary and secondary pollutants
- describe the effect of the products of combustion of hydrocarbon fuels on the environment and humans
- describe the effects of lead compounds and volatile organic compounds on the environment and humans
- explain the term 'photochemical smog'.

Primary and secondary pollutants

Atmospheric pollutants can be gases, droplets of liquid or tiny particles of solids (particulates) mixed with the air.

Primary pollutants: These are released directly from a process, e.g. carbon monoxide from car exhausts, sulphur dioxide released from burning fossil fuels, particulates released from erupting volcanoes or methane as a waste product of animal digestion. Other primary pollutants resulting from human activity include:

- nitric oxide and nitrogen dioxide from car exhausts
- carbon dioxide from lime kilns and from combustion of fossil fuels
- lead and lead compounds from combustion reactions in car engines and small particles of paints
- volatile organic compounds e.g. unburnt hydrocarbons from vehicle engines
- CFCs from refrigerants (although now banned).

Secondary pollutants: These are formed in the atmosphere when primary pollutants undergo further reactions. Secondary pollutants arising from human activity include:

- sulphur trioxide from atmospheric oxidation of sulphur dioxide
- ozone in the troposphere from the photochemical cycle involving nitrogen dioxide and oxygen (see Section 14.3 and below)
- organic compounds formed in smog (see below)
- particles of salts formed in the atmosphere, e.g. nitrates from NH_3 and acids in the air.

Burning hydrocarbon fuels

Incomplete combustion

When hydrocarbon fuels undergo complete combustion carbon dioxide is formed, which is a greenhouse gas (see Section 14.4). If not enough air is present, hydrocarbon fuels undergo incomplete combustion. Carbon monoxide, CO, is formed as well as soot (small particles of carbon).

$$2C_4H_{10}(g) + 9O_2(g) \rightarrow 8CO(g) + 10H_2O(l)$$
$$2C_4H_{10}(g) + 5O_2(g) \rightarrow 8C(s) + 10H_2O(l)$$

Particles of soot are irritants. Carbon monoxide is toxic. It combines with haem, the oxygen carrying group in the protein haemeoglobin which is present in red blood cells. It prevents oxygen from binding to haem and can lead to death.

Unburnt hydrocarbons may also be released into the atmosphere if there is not enough oxygen to burn them.

Photochemical smog

Nitrogen combines with oxygen in the high temperature of car engines to form oxides of nitrogen (NO_x). In the presence of hydrocarbons from car exhausts, ozone and sunlight, this mixture reacts to form **photochemical smog**. This is made worse in cities where a layer of warm dry air traps a

✓ Exam tips

Some pollutants can be primary as well as secondary pollutants. For example, NO_2 is a primary pollutant when emitted from car exhausts, but is a secondary pollutant when formed in the atmosphere from the reaction between nitric oxide and ozone. So when answering an exam question on primary and secondary pollutants, make sure that you understand the context in which the pollutant is being formed.

layer of cooler air beneath it (temperature inversion). The layer of warm dry air allows the maximum amount of UV light to be transmitted. Figure 14.8.1 shows how a photochemical smog forms.

Nitrogen dioxide from car exhausts can undergo photolytic reactions in the presence of UV light. Ozone is formed during this cycle and nitrogen dioxide is regenerated.

$$NO_2 \xrightarrow{\text{UV light}} NO + O\bullet$$
$$O\bullet + O_2 \longrightarrow O_3$$
$$NO + O_3 \longrightarrow NO_2 + O_2$$

When hydrocarbons and/or carbon monoxide are present, this cycle gets disrupted. Ozone reacts with these carbon compounds to produce organic radicals such as $CH_3O\bullet$ and $HCO\bullet$. These combine with nitrogen oxides to form aldehydes, peroxides and organic nitrates. These compounds can cause irritation of the eyes, breathing difficulties and asthma. Particulates may also be present, including some nitrates, aldehydes and ketones with higher molar masses. The nitrogen dioxide, which is an irritant and toxic in higher concentrations, gives the smog a brownish colour.

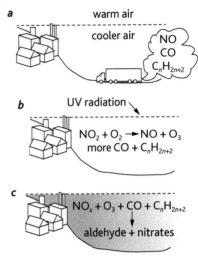

Figure 14.8.1 *The formation of photochemical smog;* **a** *Early morning: temperature inversion prevents the dispersion of NO_x, CO and hydrocarbons;* **b** *Slightly later: NO_2 reacts with O_2 to form ozone.* **c** *Late morning and afternoon: reaction of O_3, NO_x and hydrocarbons to form smog.*

Lead compounds and the environment

Lead and lead compounds are toxic. They can affect the heart, bones and kidneys. Lead in the atmosphere and in the aqueous environment is particularly harmful to the nervous system of children. It can cause permanent learning and behavioural problems. Lead gets into the environment when:

- paint containing lead is burnt
- fuel containing lead tetraethyl (added to make the fuel burn more smoothly) undergoes combustion in vehicle engines.

Organic compounds and the environment

Volatile organic compounds (VOCs) are compounds such as methane, CFCs, chlorocarbon compounds and methanal, with low boiling points. Chlorocarbons and methanal irritate the membranes of the eyes, nose and lungs, have allergic effects and can affect the immune system, especially in children. Methane is a greenhouse gas and CFCs deplete the ozone layer. VOCs get into the atmosphere:

- as combustion products of paints and glues used in building materials
- by leakage of refrigerants from old refrigerators
- by evaporation from cleaning products which contain them
- by evaporation from paints and glues.

Key points

- Primary pollutants are released directly from a process. Secondary pollutants are formed when primary pollutants react further in the air.
- Combustion of hydrocarbons may lead to global warming, the emission of CO – a toxic gas, and photochemical smog.
- Photochemical smog is formed when ozone reacts with hydrocarbons and nitrogen oxides from vehicle exhausts.
- Lead and volatile organic compounds can be released into the atmosphere when fuels burn or evaporate into the air.

On completion of this section, you should be able to:

- describe methods of controlling and preventing atmospheric pollution
- describe the importance of alternative and cleaner fuels, improved technology and mass transit in preventing pollution
- describe the importance of sequestering agents, filters, washers and scrubbers in controlling pollution.

Introduction

Many of the fuels we use, such as coal and petrol, pollute the environment and cause problems such as increased global warming and acid rain. Improved technology helps to make engine and chemical plant design more efficient so that fewer pollutants are formed. For example, electric vehicles powered by batteries do not produce carbon dioxide. **Alternative fuels** are being developed which are cleaner and reduce carbon emissions. For example, making alcohol as a fuel by fermentation is also better for the environment than the production and use of petroleum fractions (see Section 13.4).

Hydrogen is a non-polluting fuel. When it burns in oxygen, water is the only product. Hydrogen can be used as a fuel in hydrogen–oxygen **fuel cells** used to power some vehicles (Figure 14.9.1).

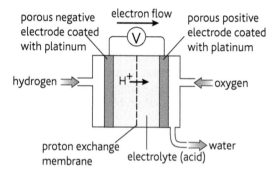

Figure 14.9.1 *A hydrogen–oxygen fuel cell*

Carbon emissions can also be reduced by:

- **mass transit** using vehicles such as buses and trains rather than individual cars
- using simpler forms of transport, such as cycles
- using alternative energy sources, e.g. solar power, wind power, wave power.

Improved technology does not solve all the problems. For example, fossil fuels may be needed to make the electricity to recharge the batteries in electric cars and to make hydrogen for fuel cells.

Catalytic converters

Catalytic converters are fitted to cars to reduce the emissions of nitrogen oxides, hydrocarbons and carbon monoxide. Once warmed up, the platinum–rhodium catalyst causes nitrogen oxides to be converted to harmless nitrogen gas and carbon monoxide to carbon dioxide. Unburnt hydrocarbons and carbon monoxide may also reduce the nitrogen oxides.

$$2NO(g) + 2CO(g) \rightarrow N_2(g) + 2CO_2(g)$$

$$2NO_2(g) + 4CO(g) \rightarrow N_2(g) + 4CO_2(g)$$

Sequestering agents

Sequestering agents are agents that remove particular ions from solution or from the air. They often form complexes with metal ions (see *Unit 1*

Study Guide, Section 13.4). They can be used to remove metals ions from the soil, from contaminated water or from the air. For example Cu^{2+} and Ni^{2+} ions can be removed from water by adding a mixture of sequestering agents such as a mixture of EDTA and chitosan. Heavy metals can be removed from industrial waste products by adding a sequestering agent then adding a surfactant such as a long-chained carboxylic acid and adjusting the pH so that the complex is uncharged. The complex can then be separated using methods which separate molecules of different solubility in polar and non-polar solvents. Ca^{2+} and Mg^{2+} ions can be removed from water by adding sodium hexametaphosphate which removes these ions as an insoluble complex.

Scrubbers

Scrubbers remove particles from waste gases in a factory. Modern scrubbers work rather like a spin dryer. The waste gases are sprayed at high pressure against the side of a cylindrical tank. The water moves up the tank in a spiral. The dirty water runs down the side of the tank and the cleaned air escapes from the top. This process is called wet scrubbing or sometimes, washing.

Flue-gas desulphurisation

Flue-gas desulphurisation is an example of scrubbing (air washing). In this process, acids, alkalis or other chemical reactions are used to neutralise harmful substances in waste gases. Sulphur dioxide can be removed from waste gases in power stations and furnaces by passing the gases through a moving bed of solid calcium carbonate or calcium oxide. The calcium sulphite formed is either dumped or made into sulphuric acid.

$$SO_2(g) + CaCO_3(s) \rightarrow CaSO_3(s) + CO_2(g)$$

$$SO_2(g) + CaO(s) \rightarrow CaSO_3(s)$$

Filters

Filters are used to remove dust and particulates from the waste gases in chemical plants and some power stations. **Air filters** consist of a number of long polyester 'socks', which allow gases through, but not dust. Air is drawn though the filters and the dust collects on the outside (Figure 14.9.2). The filter is cleaned periodically by passing air through in the opposite direction and collecting the dust as a solid.

Washing

Materials such as metal ores undergo high pressure washing to remove the contaminating dust and clays. Washing may also remove soluble contaminants from a substance.

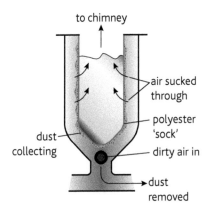

Figure 14.9.2 *A bag filter used to collect dust from waste gases*

Key points

- Pollution can be reduced by using cleaner fuels such as hydrogen or ethanol, improving technology, e.g. more efficient engines and by travelling by means of mass transport, e.g. buses and trains.

- Pollution can be reduced by using sequestering agents, filters, washers and scrubbers.

On completion of this section, you should be able to:

- understand the terms reduce, reuse and recycle
- describe the processes involved in waste reduction
- understand the importance of reusing and recycling glass, paper, plastic, steel and aluminium
- describe how to reduce the use of materials such as glass, paper, plastic, steel and aluminium.

Reduce, reuse, recycle

Reduction of waste

Waste reduction (waste minimisation) is the prevention of waste material being created. Examples are:

- Reuse of second-hand products.
- Repairing broken items instead of replacing them, e.g. mending a broken cup.
- Designing products to be reusable (using cotton shopping bags instead of plastic shopping bags).
- Avoiding the use of disposable products, e.g. disposable plastic cutlery.
- Cleaning articles before recycling.
- Designing products that use less material to achieve same purpose, e.g. lighter aluminium drinks cans or lighter steel frames with the same strength.
- Reducing excess paper or plastic packaging.
- Improving the durability of an item, e.g. making a sieve of aluminium rather than plastic.

Reuse

Reuse means to use something more than once, sometimes for the same purpose and sometimes for a different purpose. Examples are:

- Refillable drinks bottles (glass or plastic) which can be rewashed and reused.
- Using a glass jar to put flowers in.
- Retreading rubber tyres.
- Reusing metal shipping containers or wooden chests for removals.

Recycling

Recycling is the processing of used materials into fresh products. This prevents waste of potentially useful materials, reduces the consumption of fresh raw materials, reduces energy use and reduces pollution arising from incineration and landfill.

Recyclable materials can either produce a new supply of the same material (e.g. paper is converted into recycled paper) or produce a slightly different product (e.g. paperboard from paper). The disadvantages of recycling are that it requires transport to the recycling centres, the materials have to be sorted and time and energy are required for the process of recycling. It also usually produces a product of lower quality.

Examples of recycling

Glass

Waste glass can be sorted then melted and either used to make new glass objects or added as glass 'cullet' to glass being freshly made. There can be up to 30% energy saving and a 20% reduction in CO_2 emissions made by recycling glass compared with making glass from sand, lime and sodium carbonate.

Recycled glass can be used:

- to make new glass bottles or glass for display counters
- as an aggregate in concrete or in making new ceramics
- as a component of astroturf and golf-bunker 'sand'
- as an abrasive, e.g. glasspaper.

Paper

Recycled paper can come from scrap from paper mills or from household waste. Paper production accounts for 35% of the trees felled in the world, so recycling is important to reduce the number of trees cut down. Recycling paper reduces carbon emissions by about 75% and water pollution about 35% compared with making paper from trees. Recycled paper is used to make lower quality paper, cardboard and paperboard.

Plastic

Most of the plastics recycled are thermoplastics – those that melt when heated.

Recycling plastic reduces carbon emissions by about 70% compared with making plastic from their monomers. Many plastics are not recycled into the same type of plastic again. The recycled plastics formed are often not recyclable. Examples are:

- PET (poly(ethene terephthalate)) containers are melted and recycled to produce a type of polyester fibre used for fabrics, new containers, bottles, rubbish bins and plastic furniture.
- HDPE (high density poly(ethene)) is recycled to make rulers and plastic furniture.
- Poly(styrene) can be recycled to make clothes hangers, flower pots and picture frames.

Steel

Scrap iron and steel from steel plates, beams, columns, old car bodies, and cast iron pipes can be used to make new steel products such as girders. The iron or steel is melted and added to a furnace. Recycling iron and steel reduces carbon emissions by about 60% compared with making steel from haematite ore.

Did you know?

Some archaeologists have suggested that some ancient civilisations melted down scrap bronze or other metals, for example, from old axe heads, for reuse.

Aluminium

Scrap aluminium from aircraft fuselages, car bodies, cycles, cookware and aluminium cans is used to make new aluminium products such as window frames, roofing and aluminium cans. The aluminium is melted and added to a furnace then degassed to remove hydrogen. It is then recast. Recycling aluminium reduces carbon emissions by 95% and energy consumption by 95% compared with making aluminium from bauxite ore.

Key points

- Waste reduction is the prevention of waste material being created.
- Reuse means to use something more than once, sometimes for the same purpose and sometimes for a different purpose.
- Recycling is the processing of used materials into fresh products.
- Describe how to reduce the use of materials such as glass, paper, plastic, steel and aluminium.

14.11 Solid waste and the environment

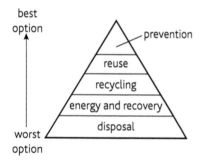

Fig 14.11.1 *Options for waste management*

Did you know?

80% of the waste in the sea is plastic. There may be as much as 10 000 million kg of plastic in the oceans.

Introduction

To avoid too much damage to the environment, we should generate the smallest amount of solid waste possible (see Section 14.10). The options for waste management are shown in Figure 14.11.1. Much of our solid waste cannot be reused or recycled but we may be able to get energy from it before it is dumped in the ground.

Incinerating waste

Some waste can be burnt in special incinerators to form a solid residue and gaseous products. This can reduce the volume of solid waste to up to one-thirtieth of its original volume. The energy released can be used to heat steam to run a turbine to produce electricity. This method is practical for disposing of some hazardous waste such as biological and medical waste. One problem with this method is that organic compounds such as dioxins, furans and polyaromatic hydrocarbons are formed. These are toxic and may stay in the atmosphere for many years. Pyrolysis of waste (heating it under pressure with limited oxygen at a high temperature) in sealed solid vessels can be used to make carbon and a gas containing CO and H_2 which is burnt to produce electricity.

Solid waste and the environment

Solid waste may harm organisms as well as the environment.

Glass: Broken glass is sharp and may cause cuts and abrasions in animals. It does not degrade very quickly. When glass breaks, tiny particles of glass are produced. These may remain in the air for some time and can cause irritation of the lungs. Pieces of glass can also act as lenses, focusing light rays and cause heating leading to fires.

Paper: Printing inks may contain heavy meals such as cadmium and arsenic as well as residual bleaches and organochlorine compounds in small quantities. These potentially toxic substances may leach into the soil when paper gets wet. In addition paper is easily blown away by the wind, so spreading litter.

Plastics: These can be a danger to wildlife, especially sea life. Animals may get tangled in plastic nets or suffocate when the plastic gets into the lungs. If plastic gets stuck in the gullet, the animal cannot ingest any food and therefore dies. Many organisms from fish to crocodiles are affected. Toxic additive may also leach out from the plastics. Since many plastics do not decompose readily and they are not biodegradable (break down by the action of microorganisms), they remain in the environment for a long time. Biodegradable plastics and other biological materials such as wood do decompose after a few years. Biodegradable plastics, however, may simply break down into microscopic particles which may be harmful to creatures in seas and rivers.

Metals: Many metals are alloys. When they are thrown away, they may react with water and air and corrode to form soluble compounds which diffuse into the soil or water. Iron rusts and may form unsightly pools of red waste which reduce plant growth as well as being unsightly. Some metals such as lead or cadmium (from car batteries) are poisonous. Waste from aluminium smelting may still contain high amounts of aluminium. This waste reacts with water forming ammonia and flammable acetylene and hydrogen.

Nuclear waste: This may contain radioactive isotopes with very long half lives, e.g. ^{129}I has a half life of 17 million years. Radioactive waste may cause radiation burns, skin damage and damage to the immune system as well as causing animals to become sterile.

Disposing of solid waste

Landfill

Waste can be buried in a landfill site such as an unused quarry or mine. Some landfills, however, consist of mounds of rubbish (waste dumps). A poorly managed site may create a number of environmental problems:

- Wind can blow away paper and plastic bags into the surrounding areas.
- Toxic or harmful liquids may drain through the soil or rocks to contaminate groundwater and soil.
- Gases such as methane, carbon dioxide and hydrogen sulphide are released as a result of organic waste breaking down in the absence of oxygen. Some of these gases are foul-smelling and may kill surface vegetation. Others (methane and CO_2) are greenhouse gases.
- Organic material may attract rats and other vermin.
- Waste dumps may take up a large area and be unstable because some of the rubbish can move.

In a well-managed landfill (sanitary landfill) site:

- The waste is compacted to prevent it moving or blowing away.
- The site has a lining of clay, plastic or rubberised material which minimises drainage of liquids into the soil or rocks below.
- The gases are extracted (the gases are either burnt off immediately or are burnt in a controlled way to generate electricity).
- The site is covered so that it does not attract rats and other vermin.
- Because it is compacted, the waste is more stable and confined to a smaller area.

Composting

Organic plant material and animal waste can be composted and returned to the soil as a fertiliser. Fungi, bacteria and worms help break down the materials in the presence of oxygen.

Nuclear waste

This requires special treatment so that the radioactive substance is completely isolated and cannot escape into ground water or air. Amongst the methods used are:

- *Vitrification:* The waste is heated then mixed with molten glass and allowed to cool. The glass does not dissolve or react with water. The glass is stored in steel cylinders in safe places underground.
- *Adsorption:* Iron(III) hydroxide or an ion-exchange resin is added to the waste to adsorb and concentrate the solution. A sludge is formed which is mixed with cement then put into drums and stored underground.
- *Above ground disposal:* (Low level waste). The waste is put into a steel cylinder. An inert gas is added. The steel cylinder is then placed in a concrete cylinder and stored.

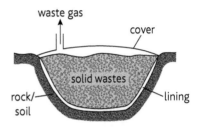

Figure 14.11.2 *A modern landfill site*

Did you know?

Nuclear waste can be reprocessed and used for other purposes. For example, ^{137}Cs and ^{90}Sr can be separated from other radioactive substances and used to irradiate food to kill bacteria.

Key points

- Iron, glass, plastic, paper and metals may all harm the environment if not disposed of correctly.
- Incineration of waste can provide energy and reduce the amount of solid going to landfill.
- A good landfill site will prevent drainage of liquid to the soil and loss of greenhouse gases to the air.
- Nuclear waste can be disposed of by vitrification or adsorption followed by encasement in cement.

Answers to all exam-style questions can be found on the accompanying CD

Multiple-choice questions

1 Which of the following half equations represents the reaction which occurs at the cathode during the electrolysis of alumina?

A $Al(l) \rightarrow Al^{3+}(aq) + 3e^-$

B $Al^{3+}(aq) + 3e^- \rightarrow Al(l)$

C $Al(l) \rightarrow Al^{3+}(l) + 3e^-$

D $Al^{3+}(l) + 3e^- \rightarrow Al(l)$

2 Which of the following are pollutants that are emitted into the atmosphere during the refining of crude oil?

i carbon monoxide

ii hydrogen sulphide

iii oxides of nitrogen

iv chlorofluorocarbons

A i and iii only

B i and iv only

C i, ii and iii only

D i, iii and iv only

3 Ammonia is manufactured during the Haber Process by combining nitrogen and hydrogen. The equation for this reaction is shown below:

$$N_2(g) + 3H_2(g) \rightleftharpoons 2NH_3(g) \quad \Delta H = -92\,kJ\,mol^{-1}$$

Which of the following represents the ideal conditions of temperature and pressure for the production of ammonia using this process?

A Low temperature and high pressure.

B High temperature and low pressure.

C Low temperature and low pressure.

D High temperature and high pressure.

4 Which of the following are necessary for the production of ethanol from the fermentation of sugars?

A Carbon dioxide, yeast and water.

B Aerobic conditions, yeast and water.

C Anaerobic conditions, yeast and water.

D Carbon dioxide, anaerobic conditions and yeast.

5 During the production of chlorine from brine using the diaphragm cell, a porous diaphragm is used. What is its function?

i To separate the liberated hydrogen and chlorine gases.

ii To separate the liberated chlorine gas from the sodium hydroxide produced.

iii To separate the liberated hydrogen from the sodium hydroxide produced.

iv To separate the brine used from the sodium hydroxide produced.

A i only

B i and ii only

C i, ii and iii only

D i, ii, iii and iv

6 During the Contact Process, sulphur trioxide is produced by reacting sulphur dioxide with oxygen as shown by the following equation:

$$2SO_2(g) + O_2(g) \rightleftharpoons 2SO_3(g) \quad \Delta H = -196\,kJ\,mol^{-1}$$

Which of the following would produce a maximum increase in the yield of sulphur trioxide?

A Low temperature, high pressure and high concentration of oxygen.

B High temperature, low pressure and low concentration of oxygen.

C Low temperature, high pressure and low concentration of oxygen.

D High temperature, low pressure and high concentration of oxygen.

7 The concentration of dissolved oxygen in water is crucial to the existence of aquatic life forms. Which of the following causes a decrease in the concentration of dissolved oxygen in water bodies?

A Low temperatures and an increase in aerobic respiration.

B High temperatures and an increase in aerobic respiration.

C Low temperatures and a decrease in aerobic respiration.

D High temperatures and a decrease in aerobic respiration.

8 A yellow precipitate was produced when a sample of river water was tested by adding iodide ions to it. Which of the following pollutants is most likely present in this sample of water?

A NO_3^- ions

B PO_4^{3-} ions

C CN^- ions

D Pb^{2+} ions

9 Which of the following species reacts with stratospheric ozone causing its depletion?

A RCl

B O_2

C $Cl\bullet$

D $ClO\bullet$

10 Fires are a major hazard when disposing of solid waste in landfills. Which of the following gases produced from landfills is responsible for these fires?

A O_2

B CO_2

C H_2

D CH_4

Structured questions

11 Aluminium is extracted by the electrolysis of molten aluminium oxide. This oxide is found in bauxite along with oxides of iron and silicon which are the main impurities.

a Write the formulae for the

i iron and

ii silicon oxides

found as impurities in aluminium oxide. [2]

b i State the acid–base nature of the oxide of silicon in **a i** above. [1]

ii Use your answer in **b i** above to explain how this oxide is removed from the bauxite, include an ionic equation in your explanation. [4]

c Explain why dissolving cryolite in the molten aluminium oxide is economically beneficial and describe how it achieves this benefit in the electrolytic process. [2]

d i Write the half equation for the reaction occurring at the anode. [2]

ii Explain why the anode must be periodically replaced. [2]

e State ONE chemical property and ONE physical property of aluminium that allows it to be used for packaging food products. [2]

12 Ammonia is manufactured during the Haber Process by combining nitrogen and hydrogen.

$$N_2(g) + 3H_2(g) \rightleftharpoons 2NH_3(g) \qquad \Delta H = -92\,kJ\,mol^{-1}$$

a How is the nitrogen for the Haber Process obtained? [1]

b i State the source of the hydrogen used in the Process. [1]

ii Write TWO equations to show how the hydrogen is obtained from the source mentioned in **b i** above. [4]

c i State the temperature and pressure that are used to manufacture ammonia using the Haber Process. [2]

ii Explain:

■ Why the temperature stated in your answer in **c i** above is considered to be a compromise temperature.

■ Why this compromise temperature is used. [2]

d Why are hydrogen and nitrogen purified before being allowed to react? [1]

e Suggest ONE reason why the ammonia is removed from the system as soon as it is formed? [2]

f State ONE use of ammonia in EACH of the following industries:

i Agriculture [1]

ii Chemical [1]

13 The chlor-alkali industry refers to the industrial production of the alkali sodium hydroxide and chlorine by the electrolysis of a concentrated solution of sodium chloride (brine). One method of production involves the use of the diaphragm cell as shown below.

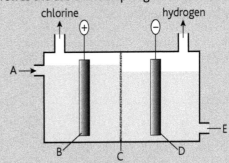

a i Complete labels A, C and E [3]

ii Describe the process by which chlorine, hydrogen and sodium hydroxide are produced in the diaphragm cell. [5]

iii Write the half equation for the reaction occurring at the anode. [2]

iv Carbon is not used to make the anode (B) and the cathode (D) in this particular cell. State the materials that the anode and cathode are made from and explain why carbon is not used. [3]

b Why is the diaphragm used in this particular cell, an environmental hazard? [1]

c State ONE use of chlorine. [1]

14 a Using balanced equations explain how the concentration of ozone is maintained in the stratosphere. [5]

b i Using the data below, describe the trend in the concentration of stratospheric ozone over time. [1]

Year	1960	1970	1980	1990	2000
[O_3]	260	240	160	110	108

ii Explain why the trend described in **b i** above is cause for concern. [2]

iii Using the relevant equations, explain why the trend described in **b i** is occurring. [4]

iv Outline ONE step that can be taken to prevent this trend from continuing. [1]

Data sheets

Selected bond energies

Diatomic molecules		Polyatomic molecules	
Bond	**Bond energy/kJ mol^{-1}**	**Bond**	**Bond energy/kJ mol^{-1}**
H—H	436	C—C	350
N≡N	994	C=C	610
O=O	496	C—H	410
F—F	158	C—Cl	340
Cl—Cl	244	C—Br	280
Br—Br	193	C—I	240
I—I	151	C—N	305
H—F	562	C—O	360
H—Cl	431	C=O	740
H—Br	366	N—H	390
H—I	299	N—N	160
		O—H	460
		O—O	150

Selected electrode potentials

Electrode reaction	$E^{\ominus}$/V
$K^+ + e^- \rightleftharpoons K$	−2.92
$Mg^{2+} + 2e^- \rightleftharpoons Mg$	−2.38
$Al^{3+} + 3e^- \rightleftharpoons Al$	−1.66
$V^{2+} + 2e^- \rightleftharpoons V$	−1.2
$Zn^{2+} + 2e^- \rightleftharpoons Zn$	−0.76
$Fe^{2+} + 2e^- \rightleftharpoons Fe$	−0.44
$V^{3+} + e^- \rightleftharpoons V^{2+}$	−0.26
$Ni^{2+} + 2e^- \rightleftharpoons Ni$	−0.25
$Sn^{2+} + 2e^- \rightleftharpoons Sn$	−0.14
$Pb^{2+} + 2e^- \rightleftharpoons Pb$	−0.13
$2H^+ + 2e^- \rightleftharpoons H_2$	0.00
$S_4O_6^{2-} + 2e^- \rightleftharpoons 2S_2O_3^{2-}$	+0.09
$Cu^{2+} + 2e^- \rightleftharpoons Cu$	+0.34
$VO^{2+} + 2H^+ + e^- \rightleftharpoons V^{3+} + H_2O$	+0.34
$I_2 + 2e^- \rightleftharpoons 2I^-$	+0.54
$Fe^{3+} + e^- \rightleftharpoons Fe^{2+}$	+0.77
$Ag^+ + e^- \rightleftharpoons Ag$	+0.80
$VO_2^+ + 2H^+ + e^- \rightleftharpoons VO^{2+} + H_2O$	+1.00
$Br_2 + 2e^- \rightleftharpoons 2Br^-$	+1.07
$Cr_2O_7^{2-} + 14H^+ + 6e^- \rightleftharpoons 2Cr^{3+} + 7H_2O$	+1.33
$Cl_2 + 2e^- \rightleftharpoons 2Cl^-$	+1.36
$MnO_4^- + 8H^+ + 5e^- \rightleftharpoons Mn^{2+} + 4H_2O$	+1.52

Key

atomic (proton) number
atomic symbol
name
relative atomic mass

IA	IIA	IIIB	IVB	VB	VIB	VIIB	VIIIB			IB	IIB	IIIA	IVA	VA	VIA	VIIA	VIIIA
1 **H** hydrogen 1.008																	2 **He** helium 4.003
3 **Li** lithium 6.941	4 **Be** beryllium 9.012											5 **B** boron 10.81	6 **C** carbon 12.01	7 **N** nitrogen 14.01	8 **O** oxygen 16.00	9 **F** fluorine 19.00	10 **Ne** neon 20.18
11 **Na** sodium 22.99	12 **Mg** magnesium 24.31											13 **Al** aluminium 26.98	14 **Si** silicon 28.09	15 **P** phosphorus 30.97	16 **S** sulphur 32.07	17 **Cl** chlorine 35.45	18 **Ar** argon 39.95
19 **K** potassium 39.10	20 **Ca** calcium 40.08	21 **Sc** scandium 44.96	22 **Ti** titanium 47.87	23 **V** vanadium 50.94	24 **Cr** chromium 52.00	25 **Mn** manganese 54.94	26 **Fe** iron 55.85	27 **Co** cobalt 58.93	28 **Ni** nickel 58.69	29 **Cu** copper 63.55	30 **Zn** zinc 65.39	31 **Ga** gallium 69.72	32 **Ge** germanium 72.61	33 **As** arsenic 74.92	34 **Se** selenium 78.96	35 **Br** bromine 79.90	36 **Kr** krypton 83.80
37 **Rb** rubidium 85.47	38 **Sr** strontium 87.62	39 **Y** yttrium 88.91	40 **Zr** zirconium 91.22	41 **Nb** niobium 92.91	42 **Mo** molybdenum 95.94	43 **Tc** technetium [98]	44 **Ru** ruthenium 101.1	45 **Rh** rhodium 102.9	46 **Pd** palladium 106.4	47 **Ag** silver 107.9	48 **Cd** cadmium 112.4	49 **In** indium 114.8	50 **Sn** tin 118.7	51 **Sb** antimony 121.8	52 **Te** tellurium 127.6	53 **I** iodine 126.9	54 **Xe** xenon 131.3
55 **Cs** caesium 132.9	56 **Ba** barium 137.3	57 **La** lanthanum 138.9	72 **Hf** hafnium 178.5	73 **Ta** tantalum 180.9	74 **W** tungsten 183.8	75 **Re** rhenium 186.2	76 **Os** osmium 190.2	77 **Ir** iridium 192.2	78 **Pt** platinum 195.1	79 **Au** gold 197.0	80 **Hg** mercury 200.6	81 **Tl** thallium 204.4	82 **Pb** lead 207.2	83 **Bi** bismuth 209.0	84 **Po** polonium [209]	85 **At** astatine [210]	86 **Rn** radon [222]
87 **Fr** francium [223]	88 **Ra** radium [226]	89 **Ac** actinium [227]	104 **Rf** rutherfordium [261]	105 **Db** dubnium [262]	106 **Sg** seaborgium [266]	107 **Bh** bohrium [264]	108 **Hs** hassium [269]	109 **Mt** meitnerium [268]									

58 **Ce** cerium 140.1	59 **Pr** praseodymium 140.9	60 **Nd** neodymium 144.2	61 **Pm** promethium [145]	62 **Sm** samarium 150.4	63 **Eu** europium 152.0	64 **Gd** gadolinium 157.3	65 **Tb** terbium 158.9	66 **Dy** dysprosium 162.5	67 **Ho** holmium 164.9	68 **Er** erbium 167.3	69 **Tm** thulium 168.9	70 **Yb** ytterbium 173.0	71 **Lu** lutetium 175.0
90 **Th** thorium 232.0	91 **Pa** protactinium [231]	92 **U** uranium 238.0	93 **Np** neptunium [237]	94 **Pu** plutonium [244]	95 **Am** americium [243]	96 **Cm** curium [247]	97 **Bk** berkelium [247]	98 **Cf** californium [251]	99 **Es** einsteinium [252]	100 **Fm** fermium [257]	101 **Md** mendelevium [258]	102 **No** nobelium [259]	103 **Lr** lawrencium [262]

Glossary

A

Absorbance The percentage (%) of light absorbed by a solution.

Accurate Accurate measurements are very close to their true values.

Acid rain Burning fossil fuels produces the acidic oxides SO_2 and NO_2, which are deposited in rainwater.

Addition reaction A single product is formed from two reactant molecules and no other product is formed.

Adsorption The process of forming bonds with a solid surface.

Air filter (industrial) Apparatus used to remove dust and particulates from waste gases in chemical plants and power stations.

Alcohol Aliphatic compound that has one or more —OH groups.

Aldehyde A compound containing a CHO functional group.

Aliphatic Compounds that contain carbon chains or branched chains.

Alkane Hydrocarbon with the general formula C_nH_{2n+2}.

Alkene Hydrocarbon that contains one or more double bonds and has the general formula C_nH_{2n}.

Alternative fuels Fuels which burn more cleanly than hydrocarbons and reduce carbon emissions.

Amide An organic compound having a $CONH_2$ functional group.

Amine An organic compound having an NH_2 functional group.

Amino acid An organic compound having an NH_2 functional group, a COOH functional group and a side chain, R, which can be acidic, alkaline or neutral.

Aromatic Compounds that have one or more cyclic rings with delocalised electrons.

Aryl compounds Compounds that have delocalised ring structure based on benzene.

Azeotropic mixture A mixture that deviates widely from Raoult's law and has a maximum or minimum boiling point. Also called a constant boiling mixture.

Azo dye The dye formed by a coupling reaction between a diazonium salt with an alkaline solution of a phenol.

B

Back titration A known amount of reagent is added in excess to the solution to be estimated. The excess reagent is then titrated.

Band region The $1300-3000\ cm^{-1}$ wavenumber region of the electromagnetic spectrum. Specific peaks in this region indicate the presence of particular groups such as C—H, O—H and C=O in the molecule.

Base peak The tallest peak in a mass spectrum.

Beer–Lambert's law The amount of light absorbed by a solution is proportional to the concentration and the path length.

Bending Mode of vibration of a molecule where one of the atoms in a bond vibrates up and down relative to the other.

Benzene The simplest aromatic hydrocarbon, C_6H_6.

Biodiesel A fuel for diesel engines made from vegetable oils or fats.

Brine A concentrated aqueous solution of sodium chloride.

C

Calibrate To correlate the readings of an instrument with those of a standard or standards.

Calibration curve A curve which relates known concentrations of a substance to a particular property, e.g. UV absorbance.

Carbocation An intermediate in organic chemistry where a carbon atom is positively charged.

Carbohydrate A molecule consisting of C, H and O where the H and O are in the ratio 2:1. The general formula for most simple carbohydrates is $C_x(H_2O)_y$.

Carbon cycle The flow of carbon through the atmosphere, living things, water and rocks which keeps the amount of carbon in the air constant.

Carboxylic acid A compound containing the COOH functional group.

Catalytic cracking Cracking using a catalyst of SiO_2 and Al_2O_3.

Catenation The ability of carbon atoms to form chains by joining.

Chain isomerism The isomers differ in the arrangement of the carbon atoms in their carbon skeleton.

Chiral centre A carbon (or other atom) with four different groups attached to it, creating the possibility of optical isomers. Some molecules, e.g. glucose, have more than one chiral centre.

Chlorofluorocarbons (CFCs) Molecules containing C, Cl and F that are responsible for depleting the ozone layer.

Chromatography Method of separating compounds by differences in solubility and/or charge.

cis-trans isomerism See geometric isomerism.

Component One of the compounds or elements in a mixture.

Condensation reaction When two molecules react and a small molecule is eliminated (given off).

Condensed formula A formula of an organic compound showing each carbon atom and the atoms attached but not the bonds.

Conductimetric titrations Titrations involving measurement of changes in electrical conductivity.

Conjugative effect Occurs in some molecules with multiple bonds, e.g. benzene. The effect makes the bond lengths and bond strengths intermediate between ordinary single bonds and ordinary double bonds.

Constant boiling mixture See azeotropic mixture.

Contact Process Sulphuric acid is made by this process, which refers to the conversion of SO_2 to SO_3 using a V_2O_5 catalyst and a temperature of 450 °C.

Coupling reaction The reaction of a diazonium salt with an alkaline solution of a phenol to form an azo dye.

Cracking The thermal decomposition of alkanes into shorter-chain alkanes and alkenes.

Cryolite Compound used to dissolve alumina and lower its melting point during the electrolysis of alumina.

D

Dehydration A reaction in which water is removed (eliminated).

Delocalised Electrons whose orbitals can extend over three or more atoms, allowing movement of electrons over more than two atoms.

Denitrification The reduction of nitrates to N_2 gas by bacteria.

Desalination The removal of salts from water, usually from seawater.

Diazonium salt A salt of general formula $RN^+\equiv NX^-$.

Diazotisation The formation of a diazonium salt by the reaction of an aromatic amine with nitrous acid.

Displayed formula A formula showing all the atoms and bonds.

Distribution coefficient The equilibrium constant for the distribution of solvent between two immiscible solutes.

E

Electrodialysis Method of desalination where ions are transported from one solution to another, using an ion-exchange membrane.

Electromagnetic radiation Waves which have electrical and magnetic components.

Electrophile A positively charged or partially positively charged reagent that attacks an electron-rich area of a molecule.

Empirical formula Formula showing the simplest ratio of atoms of each element in the compound.

Ester Compound containing the

$$-\overset{|}{\underset{|}{C}}-O-\overset{O}{\overset{\|}{C}}-\overset{|}{\underset{|}{C}}-\text{ functional group.}$$

Esterification The reaction of an alcohol with a carboxylic acid to make an ester.

Eutrophication The pollution of rivers by fertilisers leading to the death of plants and animals.

F

Fehling's solution A solution used to distinguish aldehydes from ketones.

Fermentation (alcoholic) Making ethanol from sugars by using yeast under anaerobic conditions (no oxygen present).

Filter (industrial) See air filter.

Fingerprint region The 600–1300 cm^{-1} wavenumber region of the electromagnetic spectrum. Peaks in this region tell us about the structure of the whole molecule.

Fraction A group of compounds that separate from a mixture within a narrow range of boiling points.

Fractional distillation A process used to separate mixtures of liquids of slightly different boiling points.

Fragmentation The breakdown of a compound in a particular way in a mass spectrometer.

Free radical Atoms or groups of atoms with an unpaired electron.

Frequency The number of waves passing a given point per second.

Fuel cell An electrochemical cell in which O_2 and H_2 react to produce water.

Functional group An atom or group of atoms that give a compound its particular chemical properties.

Functional group isomerism The molecular formula of the isomers is the same but the functional groups are different.

G

Gas–liquid chromatography Chromatography in which the stationary phase is a liquid and the mobile phase is a gas.

Geometrical isomerism Two substituent groups either side of a double bond are arranged either on the same side (*cis*) or on the opposite sides (*trans*).

Global warming The rise in temperature of the atmosphere due to the greenhouse effect.

Gravimetric analysis Determining the amount of a substance present in a compound by methods involving weighing.

Greenhouse effect The process by which thermal radiation is absorbed by the atmosphere and re-radiated in all directions.

Greenhouse gas A gas that absorbs and emits infrared radiation.

H

Haber Process The process for making ammonia from N_2 and H_2 using an iron catalyst.

Halogenoalkane Alkane in which one or more H atoms are substituted by halogen atoms.

Heterolytic fission The breaking of a bond so that the two shared electrons in the bond are split unequally between the two atoms. One of the atoms keeps both the electrons and so becomes negatively charged. The other atom becomes positively charged.

Homologous series A group of organic compounds with the same functional group in which each successive member increases by a CH_2 unit.

Homolytic fission The breaking of a bond so that the two shared electrons in the bond are split equally between the two atoms, one electron going to each atom.

Hybridisation The process of mixing atomic orbitals.

Hydrocarbons Compounds containing carbon and hydrogen only.

Hydrolysis The breakdown of a compound with water. The rate of reaction is often increased by reacting the compound with an acid or an alkali.

I

Ideal solution A solution which obeys Raoult's law.

Immiscible liquids Liquids which do not dissolve in each other.

Inductive effect The ability of groups of atoms to exert an electron-attracting or withdrawing effect on the electrons around a particular atom.

Infrared (IR) Radiation of wavelength about 700–10^6 nm.

Initiation The first step in a photochemical reaction in which free radicals are formed.

Iodoform reaction Compounds containing the CH_3CHOH group are oxidised by I_2 and NaOH to form a yellow precipitate of triiodomethane.

Isomers Molecules that have the same molecular formula but the atoms are arranged differently.

K

Ketone A compound containing a CO functional group between two carbon atoms.

M

$M+1$ peak Small peak in a mass spectrometer trace one m/z unit beyond the molecular ion peak.

Mass/charge ratio In mass spectrometry, the mass of an ion divided by its charge m/z.

Mass spectrometer Instrument used to calculate relative atomic masses and to identify organic compounds.

Mass transit Using vehicles such as buses and trains for transportation rather than individual cars.

Mean The average of the numbers taken from the data in identical experiments.

Mesomerism Making up a composite structure from several different structures.

Miscible liquids Liquids able to mix with one another.

Mobile phase The phase that moves over the stationary phase in chromatography.

Molecular formula A formula showing the actual number of atoms of each element present in a molecule of a compound.

Molecular ion peak The peak in a mass spectrum arising from the removal of one electron from a molecule.

Monomer Small molecules that join together to form polymers.

N

Nitration In organic chemistry, the substitution of a H atom by an NO_2 group.

Nitrogen cycle The flow of nitrogen through the atmosphere, living things, water and rocks that keeps the amount of N_2 in the air constant.

Nitrogen fixation The conversion of N_2 in the air to ammonia by bacteria.

Nucleophile A reagent that donates a pair of electrons to an electron-deficient atom in a molecule.

Nucleophilic substitution A reaction in which the nucleophile bonds with or 'attacks' the positive or partially positive charge of an atom in a molecule (usually a carbon atom) resulting in the replacement of a group attached to it.

O

Optical isomerism This occurs when four different groups are attached to a central carbon atom. The two isomers formed are mirror images of each other.

Ozone depletion The decrease in the amount of ozone in the ozone layer caused by CFCs.

Ozone layer An area within the stratosphere that has a relatively high concentration of ozone.

P

Partial vapour pressure The pressure exerted by each component in the vapour alone.

Partition coefficient See distribution coefficient.

Partitioning Dividing the components of a mixture between two different phases.

Phenol Compound containing one or more —OH groups attached directly to an aromatic ring.

Photochemical smog A smoky fog formed when hydrocarbons, nitrogen oxides and ozone react in the presence of UV light.

Photodissociation The breaking of a bond by light (usually UV light).

Polyamide Polymers with many amide linkages, —CONH—.

Polyester Polymers with many ester linkages, —COO—.

Polymer Large molecule built up from many small molecules.

Polymerisation The process of forming polymers from monomers.

Polysaccharide Polymer of simple sugar units.

Positional isomerism The position of the functional group is different but the molecular formula is the same.

Potentiometric titration Titration involving measurement of changes in electrode potentials.

Precise (precision) Precise measurements are very close to each other in value.

Primary (referring to alcohols and halogenoalkanes) The OH or Cl is attached to a carbon atom, which is attached to only one other carbon atom.

Primary pollutant Pollutant released directly from a process.

Primary standard A chemical whose properties make it suitable for deducing the concentration of other acids and alkalis.

Propagation A cyclic series of reactions in which free radicals react with molecules or atoms to form different radicals and different molecules or atoms.

Proteins Natural polymers made from 20 naturally occurring amino acids.

Q

Quanta Energies of fixed values only, which can be absorbed or emitted by an atom.

R

Radical See free radical.

Raoult's law The partial vapour pressure of a component in a mixture = its mole fraction × vapour pressure of pure component.

Reaction mechanisms These show the steps in bond breaking and bond making when reactants are converted to intermediates and then to products.

Recycling The processing of used materials into new products.

Redox titration These are used to calculate the concentration of oxidising or reducing agents.

Reforestation Planting of young trees to replace depleted forests or woodland. Replenishes the wood resource and absorbs more CO_2 from the air.

Reforming The conversion of alkanes to cycloalkanes or cycloalkanes to arenes.

Relative abundance The relative amount of one species compared with the commonest species in a mass spectrum.

Repeating unit The smallest group of atoms in a polymer derived from a monomer, which when joined gives the structure of the polymer.

Resonance hybrid The composite structure of a molecule made up of a synthesis of several different forms.

Retention time In gas chromatography, the time between injection of a compound and its detection.

Retention value, R_f In chromatography the distance moved by a compound from the base line, divided by the distance of the solvent front from the base line.

Reuse To use something more than once for the same or a different purpose.

Reverse osmosis Method of desalination in which water is forced through a semi-permeable membrane from a region of high to low salt concentration.

S

Saponification The process of making soaps by the hydrolysis of fats and oils.

Saturated Compounds that contain only single bonds so that no more hydrogen can be added.

Scrubber Part of a chemical plant that removes particles from waste gases using a spray of water.

Secondary (referring to alcohols and halogenoalkanes) The OH or Cl is attached to a carbon atom which is attached to two other carbon atoms.

Secondary pollutant Pollutant formed when primary pollutants undergo further reactions.

Sequestering agent Agent which removes particular ions from solution or from the air.

Solvent extraction The separation of a solute because of differences in its solubility in two solvents.

Solvent front The leading edge of the solvent that progresses along the surface where the separation of the mixture (chromatography) is occurring.

Standard deviation A measure of how spread out the data is from the mean.

Stationary phase In chromatography, a solid or liquid that remains fixed in position.

Steam distillation Distillation of a more volatile compound from an immiscible mixture using steam.

Stereoisomerism Two compounds have the same atoms bonded to each other but the atoms have a different arrangement in space.

Stretching Mode of vibration of a molecule where the atoms in a bond vibrate in one plane.

Structural formula A formula showing the arrangement of atoms in a molecule in a simplified form.

Structural isomers Compounds with the same molecular formula but different structural formulae.

Substitution reaction A reaction in which one atom or group of atoms is replaced by another.

T

Termination A reaction in which two free radicals combine to form a molecule.

Tertiary (referring to alcohols and halogenoalkanes) The OH or Cl is attached to a carbon atom which is attached to three other carbon atoms.

Tetravalency The atom has four valence electrons in its outer principal quantum shell. These form four bonds with other atoms.

Thermometric titrations Titrations involving measurement of changes in temperature.

Thin-layer chromatography A form of chromatography in which the mobile phase is a liquid and the stationary phase a thin layer of solid.

Tollens' reagent Ammoniacal silver nitrate used for the silver mirror test to distinguish aldehydes from ketones.

Transesterification The reaction of an ester with an alcohol to form a different ester and a different alcohol.

Transmission The percentage (%) of light passing through a solution.

Turbidity The cloudiness of suspended matter in a liquid.

U

Ultraviolet (UV) Radiation of wavelength between about 4–400 nm.

Unsaturated Compounds containing double or triple bonds (usually carbon–carbon bonds). Hydrogen can be added to these compounds to make them saturated.

V

Vacuum distillation Distillation under reduced pressure.

Vapour pressure The pressure exerted by vapour molecules in a closed system.

Visible (vis) Radiation of wavelength about 400–700 nm.

Visualising agent Chemical used in chromatography to make colourless spots coloured.

W

Waste reduction The prevention of waste material being created.

Water cycle The process in which the water on the surface of the Earth and in the atmosphere is constantly evaporating and condensing.

Wavenumber In IR spectroscopy, the frequency of vibration divided by the speed of light.

Z

Zwitterion An electrically neutral species with a positive and negative charge in two different parts of the ion.

Index